Developing Modern Database Applications with PostgreSQL

Use the highly available and object-relational PostgreSQL to build scalable and reliable apps

Dr. Quan Ha Le
Marcelo Diaz

BIRMINGHAM - MUMBAI

Developing Modern Database Applications with PostgreSQL

Group Product Manager: Kunal Parikh
Publishing Product Manager: Ali Abidi
Senior Editor: Mohammed Yusuf Imaratwale
Content Development Editor: Joseph Sunil
Technical Editor: Sonam Pandey
Copy Editor: Safis Editing
Project Coordinator: Aparna Ravikumar Nair
Proofreader: Safis Editing
Indexer: Vinayak Purushotham
Production Designer: Alishon Mendonca

First published: Aug 2021

Production reference: 2190821

Published by Packt Publishing Ltd.
Livery Place
35 Livery Street
Birmingham
B3 2PB, UK.

ISBN 978-1-83864-814-5

www.packt.com

About the authors

Dr. Quan Ha Le graduated with a Ph.D. in computer science from the Queen's University of Belfast, United Kingdom, in 2005. Since his Ph.D. graduation, he has been working as a PostgreSQL database administrator all over Alberta and Ontario, Canada, until now.

From 2008 to 2019, Dr. Le Quan Ha administered, designed, and developed 24 small, medium, large, and huge PostgreSQL databases in Canada. Since 2016, after writing a good publication on PostgreSQL database clusters on clouds, he has been a member of the **United States PostgreSQL Association** (**PgUS**) in New York City. Dr. Le Quan Ha has also been a board member of the PgUS Diversity committee since 2018.

I want to thank the people who have been close to me and supported me, especially the Editors - Gebin George, Sonam Pandey, Aishwarya Mohan, Sonam Pandey, Joseph Sunil, and Priyanka Mhatre.

Marcelo Diaz is a software engineer with more than 15 years of experience, with a special focus on PostgreSQL. He is passionate about open source software and has promoted its application in critical and high-demand environments where he has worked as a software developer and consultant for both private and public companies. He currently works very happily at Cybertec and as a technical reviewer for Packt Publishing. He enjoys spending his leisure time with his daughter, Malvina, and his wife, Romina. He also likes playing football.

To my family for supporting especially those long hours after normal work and weekends.

About the reviewer

Enrico Pirozzi has been a PostgreSQL **Database Administrator (DBA)** since 2003 , and has an EnterpriseDB certification on implementation management and tuning, along with a master's in computer science. Based in Italy, he has been providing database advice to clients in industries such as manufacturing and web development for over 20 years. Dedicated to open source technology since early in his career, he is a co-founder of the PostgreSQL Italian mailing list, PostgreSQL-it, and of the PostgreSQL Italian community site, PSQL.it. In 2018, he wrote *PostgreSQL 10 High Performance* and in 2020, he wrote *Learn PostgreSQL*, both available from Packt Publishing. He now works at Nexteam ad DBA Postgresql, a nice place where a PostgreSQL DBA can live and work.

Packt is searching for authors like you

If you're interested in becoming an author for Packt, please visit authors.packtpub.com and apply today. We have worked with thousands of developers and tech professionals, just like you, to help them share their insight with the global tech community. You can make a general application, apply for a specific hot topic that we are recruiting an author for, or submit your own idea.

`Packt.com`

Subscribe to our online digital library for full access to over 7,000 books and videos, as well as industry leading tools to help you plan your personal development and advance your career. For more information, please visit our website.

Why subscribe?

- Spend less time learning and more time coding with practical eBooks and Videos from over 4,000 industry professionals

- Improve your learning with Skill Plans built especially for you

- Get a free eBook or video every month

- Fully searchable for easy access to vital information

- Copy and paste, print, and bookmark content

Did you know that Packt offers eBook versions of every book published, with PDF and ePub files available? You can upgrade to the eBook version at `www.packt.com` and as a print book customer, you are entitled to a discount on the eBook copy. Get in touch with us at `customercare@packtpub.com` for more details.

At `www.packt.com`, you can also read a collection of free technical articles, sign up for a range of free newsletters, and receive exclusive discounts and offers on Packt books and eBooks.

Table of Contents

Preface

PostgreSQL is an open source object-relational **Database Management System** (**DBMS**) that provides enterprise-level services, including high performance and scalability. This book is a collection of unique projects providing you with a wealth of information relating to administering, monitoring, and testing PostgreSQL. The focus of each project is on both the development and administrative aspects of PostgreSQL.

Starting by exploring development aspects such as database design and its implementation, you'll then cover PostgreSQL administration by understanding PostgreSQL architecture, PostgreSQL performance, and high-availability clusters. Various PostgreSQL projects are explained through current technologies such as DevOps and cloud platforms using programming languages such as Python and Node.js. Later, you'll get to grips with the well-known database API tool, PostgREST, before learning how to use popular PostgreSQL database testing frameworks. The book is also packed with essential tips and tricks and common patterns for working seamlessly in a production environment. All the chapters will be explained with the help of a real-world case study on a small banking application for managing ATM locations in a city.

By the end of this DBMS book, you'll be proficient in building reliable database solutions to meet your organization's needs.

Who this book is for

This PostgreSQL book is for database developers, database administrators, data architects, or anyone who wants to build end-to-end database projects using Postgres. This book will also appeal to software engineers, IT technicians, computer science researchers, and university students who are interested in database development and administration. Some familiarity with PostgreSQL and Linux is required to grasp the concepts covered in the book effectively.

What this book covers

Chapter 1, *Introduction to PostgreSQL Development and Administration*, introduces the development of PostgreSQL and how PostgreSQL has become a popular **Database as a Service (DBaaS)** among the current clouds. We will present an overview of the numerous features of PostgreSQL in various development environments, such as Node.js and Django. Moving forward, we will introduce you to the PostGIS extension, which is a PostgreSQL facility for a geospatial PostgreSQL database. We will also present the PostgREST standalone web server, which aims to do one thing well: add an HTTP interface to any PostgreSQL databases or RESTful APIs. Then, in the second part of the chapter, we will learn about the administration of PostgreSQL.

Chapter 2, *Setting Up a PostgreSQL RDS for ATM Machines*, describes the DBaaS options for PostgreSQL through Amazon RDS for PostgreSQL.

Chapter 3, *Using PostgreSQL and Node.js for Banking Transactions*, describes the steps to create a RESTful web service that will be an **Application Programming Interface (API)** based on Node.js, Express, and PostgreSQL and will implement HTTP requests such as the GET, PUT, POST, and DELETE methods on data.

Chapter 4, *Managing Bank ATM Locations Using PostgreSQL and Django*, shows how to install and configure PostgreSQL so that you can use it with a Django application.

Chapter 5, *Creating a Geospatial Database Using PostGIS and PostgreSQL*, introduces you to PostGIS, a spatial extension for PostgreSQL. Through the project in the chapter, we will learn to implement geographic mapping for our Postgres **Relational Database Service (RDS)**.

Chapter 6, *Managing Banking Transactions Using PostgREST*, teaches how to develop with PostgREST. The project in this chapter will help us use PostgREST to create an API for sending requests to the RDS.

Chapter 7, *PostgreSQL with DevOps for Continuous Delivery*, teaches how to set up DevOps tasks for PostgreSQL databases.

Chapter 8, *PostgreSQL High Availability Clusters*, focuses on **High Availability (HA)** clusters and how to achieve them with PostgreSQL.

Chapter 9, *High-Performance Team Dashboard Using PostgreSQL and New Relic*, demonstrates how to install and activate PostgreSQL integration and will help you gain an understanding of the data collected by the New Relic infrastructure.

`Chapter 10`, *Testing the Performance of Our Banking App with PGBench and JMeter*, shows us how to create a load test for a PostgreSQL database to benchmark PostgreSQL performance with PGBench and JMeter.

`Chapter 11`, *Test Frameworks for PostgreSQL*, showcases how to write automated tests for existing stored procedures and develop procedures using the concepts of unit tests and **Test-Driven Development** (**TDD**).

`Chapter 12`, *Appendix - PostgreSQL among the Other Current Clouds*, talks about the various DBaaS options for PostgreSQL with other cloud platforms.

To get the most out of this book

Software/hardware covered in the book	OS requirements
PostgreSQL 11 and above	Windows/macOS/Linux

If you are using the digital version of this book, we advise you to type the code yourself or access the code via the GitHub repository (link available in the next section). Doing so will help you avoid any potential errors related to the copying and pasting of code.

Download the example code files

You can download the example code files for this book from GitHub at `https://github.com/PacktPublishing/Developing-Modern-Database-Applications-with-PostgreSQL`. In case there's an update to the code, it will be updated on the existing GitHub repository.

We also have other code bundles from our rich catalog of books and videos available at `https://github.com/PacktPublishing/`. Check them out!

Download the color images

We also provide a PDF file that has color images of the screenshots/diagrams used in this book. You can download it here: `https://static.packt-cdn.com/downloads/9781838648145_ColorImages.pdf`

Conventions used

There are a number of text conventions used throughout this book.

`CodeInText`: Indicates code words in text, database table names, folder names, filenames, file extensions, pathnames, dummy URLs, user input, and Twitter handles. Here is an example: "This `getAllATMLocations` module will first send a `SELECT` statement to PostgreSQL to retrieve all of the ATM locations."

A block of code is set as follows:

```
// add query functions
module.exports = {
    getAllATMLocations: getAllATMLocations,
    getSingleATMLocation: getSingleATMLocation,
    createATMLocation: createATMLocation,
    updateATMLocation: updateATMLocation,
    removeATMLocation: removeATMLocation
};
```

When we wish to draw your attention to a particular part of a code block, the relevant lines or items are set in bold:

```
[default]
exten => s,1,Dial(Zap/1|30)
exten => s,2,Voicemail(u100)
exten => s,102,Voicemail(b100)
exten => i,1,Voicemail(s0)
```

Any command-line input or output is written as follows:

```
[centos@ip-172-31-95-213 ~]$ sudo su
[root@ip-172-31-95-213 centos]# cd /usr/local/src/
```

Bold: Indicates a new term, an important word, or words that you see onscreen. For example, words in menus or dialog boxes appear in the text like this. Here is an example: "Right-click on **PostGIS** in the list under the **Browser** panel as shown in the following screenshot and select **New Connection**:"

Note that a base box is a template of a virtual machine, defined on the Vagrant site as follows: "Boxes are the package format for Vagrant environments. A box can be used by anyone on any platform that Vagrant supports to bring up an identical working environment."

 Tips and tricks appear like this.

Get in touch

Feedback from our readers is always welcome.

General feedback: If you have questions about any aspect of this book, mention the book title in the subject of your message and email us at customercare@packtpub.com.

Errata: Although we have taken every care to ensure the accuracy of our content, mistakes do happen. If you have found a mistake in this book, we would be grateful if you would report this to us. Please visit www.packtpub.com/support/errata, selecting your book, clicking on the Errata Submission Form link, and entering the details.

Piracy: If you come across any illegal copies of our works in any form on the Internet, we would be grateful if you would provide us with the location address or website name. Please contact us at copyright@packt.com with a link to the material.

If you are interested in becoming an author: If there is a topic that you have expertise in and you are interested in either writing or contributing to a book, please visit authors.packtpub.com.

Reviews

Please leave a review. Once you have read and used this book, why not leave a review on the site that you purchased it from? Potential readers can then see and use your unbiased opinion to make purchase decisions, we at Packt can understand what you think about our products, and our authors can see your feedback on their book. Thank you!

For more information about Packt, please visit packt.com.

Section 1 - Introducing PostgreSQL Development and Administration

This section provides a brief introduction to the topics covered in the book, namely development and administration in PostgreSQL. This section contains the following chapter:

- Chapter 1, *Introduction to PostgreSQL Development and Administration*

1
Introduction to PostgreSQL Development and Administration

PostgreSQL is an **object-relational database management system** (**ORDBMS**) based on the INGRES (**IN**teractive **G**raphics **RE**trieval **S**ystem) package, which was developed at the University of California, Berkeley. The POSTGRES (**Post** In**gres**) project started in 1985, and version 1 was released to a small number of external users in June of 1989. Now, with more than 20 years of development, PostgreSQL has become the most advanced open source database, available all over the world.

This chapter introduces the development of PostgreSQL and how PostgreSQL has become a popular **Database as a Service** (**DBaaS**) among the current clouds. We will present an overview of the numerous features of PostgreSQL in various development environments, such as NodeJS and Django. Moving forward, we will introduce you to the PostGIS extension, which is a PostgreSQL facility for a geospatial PostgreSQL database. We will also present the PostgREST standalone web server, which aims to do one thing well: add an HTTP interface to any PostgreSQL databases or RESTful APIs.

Then, in the second part of the chapter, we will learn about the administration of PostgreSQL. We will utilize DevOps through the setting up of PostgreSQL **high availability** (**HA**) clusters. We will also set up New Relic to monitor a PostgreSQL database, carry out performance tests on a PostgreSQL database with PGBench and JMeter, and use PostgreSQL testing frameworks.

In this chapter, we will cover the following main topics:

- An overview of PostgreSQL development
- An overview of PostgreSQL administration

An overview of PostgreSQL development

In 1994, Postgres95 was released to the world by Andrew Yu and Jolly Chen as an open source descendant of the original POSTGRES Berkeley code; they added a SQL language interpreter to POSTGRES. By 1996, the name "Postgres95" was changed to a new name – PostgreSQL Version 6.0 – combining the original POSTGRES version and recent versions with SQL capability. Recently, *DB-Engines* and the *SD Times 2018 100* have featured PostgreSQL as the "DBMS of the Year 2017."

In October 2019, the first PostgreSQL 12.0 version was released by the PostgreSQL Global Development Group. Since then, PostgreSQL has been the most advanced open source database all over the world. Version 12 provides many important improvements, including the following:

- The increased performance of standard B-tree indexes and also the reduction of the index size for B-tree indexes.
- The ability to rebuild indexes concurrently so that a REINDEX operation will not block any index writes; the parallel indexes introduced from PostgreSQL 10 now get more benefits with the new REINDEX CONCURRENTLY statement.
- The efficient capabilities regarding partitioning performance so that developers can now process (that is, query or alter) thousands of partitions simultaneously without blocking, and they can use foreign keys to reference partitioned tables.
- The most common value statistics for the CREATE STATISTICS command, leading to improved query plans.
- The common table expressions using WITH queries can now be inlined for quicker queries.
- The INCLUDE clause for **generalized search tree** (**GiST**) indexes is an extensible data structure that allows you to develop indices over any kind of data.
- The reduction of **write-ahead log** (**WAL**) overheads generated from a GiST, GIN, or SP-GiST index.
- The checksum control ability via the pg checksums statement (which used to be pg_verify_checksums) so that developers can enable or disable an offline cluster without dumping and reloading data (note that online checksum enablement is still in progress and not yet available in PostgreSQL 12).

Since October 2019, the latest PostgreSQL version 12 has been released with more improvements to the performance of the INSERT and COPY statements for partitioned tables and the attachment of a new table partition without blocking queries. You can read more about the fixes to previous releases at https://www.postgresql.org/docs/12/release-12.html.

Finally, PostgreSQL 12 benefits all users with notable improvements to query performance over larger datasets and space utilization; PostgreSQL 12 has quickly received good reviews and evaluations across the database industry. Version 12 of PostgreSQL is obviously the preferred open source database for all developers.

PostgreSQL 12 is now developed on clouds or so-called cloud databases and DBaaS.

What is DBaaS?

DBaaS, which is sometimes referred to as a cloud database, provides many kinds of databases as a managed service. DBaaS works in the same way as **Infrastructure as a Service (IaaS)** or **Platform as a Service (PaaS)**. IaaS provides infrastructure components and PaaS provides development platforms as managed services in the cloud. In fact, the offerings of IaaS and PaaS often include databases.

When customers demand a DBaaS offering in the cloud, they only pay for what they use on a monthly or annual basis. They do not have to pay for what they do not use. The cloud providers are responsible for managing database services such as maintenance, upgrades, or administration for their customers. At present, the DBaaS offerings include both **relational database management systems** (**RDBMS**) and NoSQL databases.

The primary advantage of the cloud service model is that customers do not have to install or maintain their software in the data center; it is well understood by every developer. However, there are also some disadvantages, such as a lack of control over network performance issues or the inability to compress data or other database maintenance tasks.

PostgreSQL can be delivered as DBaaS on many clouds, such as **Amazon Web Services** (**AWS**), Google Cloud SQL, Microsoft Azure, Heroku, and EnterpriseDB Cloud.

The development of PostgreSQL by various environments

Most applications, at some point, need to persist data. This can be through files, local storage, cloud services, or often databases. Relational database systems are usually a good default choice, particularly PostgreSQL, which is a very powerful open source SQL server.

Some companies have resources to organize their own dedicated database team. If you are lucky enough to work in such a company, they will probably craft all of their stored procedures inside their databases. So, you will only have to use the language of your choice – NodeJS or Python – to call these stored procedures. In practice, this might not occur very often, and the truth is that many developers, or even so-called full stack developers, know very little about SQL and will seek out various abstractions in their favorite package repository.

Features of PostgreSQL with NodeJS

As a newfangled piece of technology, NodeJS is a cutting-edge web server that generates vast attention, not only from start-ups but also from giant enterprises. NodeJS is considered an exceptional framework for the IT market because of the following factors:

- NodeJS is an open source cross-platform that is built on the V8 JavaScript runtime of Chrome.
- NodeJS is ideal for data-intensive, real-time applications, and it enables event-driven programming.
- NodeJS is also a super-fast JavaScript interpreter because it works on a non-blocking I/O model.
- Outperforming conventional web servers, NodeJS has been a viable option for Yahoo, eBay, and Walmart.
- NodeJS employs push technology.
- It is also a lightweight web server in terms of memory usage.
- NodeJS can be a good solution for data-dense, real-time web applications across multiple devices.

The following features of NodeJS are usually emphasized by developers:

- **The super-fast web server**: NodeJS operates at a very high speed by using a single-threaded event loop model to process all of its asynchronous I/O operations. Therefore, any major actions can be performed quickly with NodeJS, such as network connections, filesystems, and reading/writing to databases. NodeJS supports developers by allowing them to create quick and robust network applications and offers parallel connections to increase throughput.

- **The JavaScript web server**: The NodeJS suite is a JavaScript runtime environment, and developers can now write JavaScript not only in the browser but also on the server. When the browser code and the server code are created in a similar manner, it is convenient to transport data between the server and the client. NodeJS fills the gap between the frontend and backend skills, and full stack developers can use JavaScript for both. The fact that all NodeJS programs are made using JavaScript increases the effectiveness of the web development process.
- **Real-time data streaming**: NodeJS considers both HTTP requests and responses as data streams. Hence, when data comes in the form of streams, the overall processing time can be reduced because NodeJS can process files and upload files at the same time. Developers can choose NodeJS for real-time video or audio recording.
- **Real-time web applications**: Because NodeJS is very fast, it is obviously a winner for games and chat apps as well as any other multi-user real-time web apps. The synchronization process is fast and orderly due to the event-driven architecture of NodeJS, and the event loop of NodeJS through the web socket protocol handles the multi-user function.
- **Node Package Manager (NPM)**: NodeJS is an open source suite with more than 60,000 modules in the NPM.
- **A good solution for synchronization**: NodeJS is very efficient in its ability to solve common network development problems because it manages asynchronous I/O very well with its single-threaded event loop. In order to handle many clients, all I/O tasks are undertaken together in NodeJS.

Features of PostgreSQL with Python

Many Python developers prefer to use PostgreSQL for their applications. Therefore, Python's sample code for using PostgreSQL has been better documented with typical usage scenarios. Both PostgreSQL and Python are popular because of the following:

- PostgreSQL's open source license is free, so developers can easily operate as many databases as they wish without any cost.
- We can now find developers who have PostgreSQL experience more easily than other relational databases.
- Django is a Python **Model-View-Template** (**MVT**) web framework, which defines itself as a "batteries included" web framework.
- Django is simple and robust, and it is one of the most famous frameworks used by websites such as Instagram, YouTube, Google, and NASA.

The following features are emphasized by developers of Python (Django):

- **PostgreSQL's connection with Python**: Python developers connect to PostgreSQL databases via the `psycopg2` database driver and then use an **object-relational mapper** (**ORM**) to turn PostgreSQL tables into objects; finally, these objects can be utilized in their Python applications.
- **A "batteries included" philosophy**: The philosophy of "batteries included" means that Python provides plenty of functionalities, including magical ORM, multisite and multi-language support, MVT layouts, RSS and Atom feeds, AJAX support, free APIs, URL routing, easy database migrations, session handling, HTTP libraries and templating libraries, code layouts, and a default admin section.
- **Good tutorials**: Django has very good documentation that developers can quickly update with the latest information, such as technical requirements and quick-start details, detailed release notes, backward-incompatible changes, and online topics on Python development.
- **The Django admin interface**: The admin interface is one of the key advantages of Django. From a few simple lines of Python code, developers can get a good featured admin interface.
- **Built-in templates**: One of the advantages of Django is its robust built-in template system that facilitates the process of application development.
- **Good scalability**: Django can handle heavy traffic and the mobile app usage of many users; it maximizes scalability while minimizing web hosting costs. Django can also execute fine for a large number of hosts while maintaining a relatively cheap or even free hosting price.
- **Best web application security**: Django has protection against SQL injections, XSS and CSRF attacks, and clickjacking, as it hides the source code of websites.

PostGIS spatial extension

PostGIS is a spatial extender for PostgreSQL. It can be a good *OpenGIS Simple Features for SQL*-compliant spatial database because it allows you to use location SQL queries for geographic objects while remaining free and open source. Because spatial data is usually related to various types of data, PostGIS allows PostgreSQL developers to encode more complex spatial relationships. The first version (0.1) of PostGIS was released in May 2001 by Refractions Research. The latest released version of PostGIS is PostGIS 3.1.2.

When using PostGIS, we are able to execute our spatial data just like anything else in SQL statements. With the power of SQL, developers can conveniently perform spatial database transactions, backups, integrity checks, less data redundancy, multi-user operations, and security controls.

The following is a list of advantages of PostGIS:

- It offers complicated spatial tasks, spatial operators, and spatial functions.
- It significantly shortens the development time of applications.
- It allows spatial SQL querying using simple expressions for spatial relationships, such as the distance, adjacency, and containment, and for spatial operations, such as the area, length, intersection, union, and buffer.

In 2006, the Open Geospatial Consortium evaluated that "PostGIS implements the specified standard for simple features for SQL." In fact, PostGIS can be used as a backend for many software systems such as OpenStreetMap, ArcGIS, OpenJUMP, MapGuide, Kosmo, and QGIS. Furthermore, PostGIS also has an open source extension named "pgRouting" that provides geospatial routing functionality with many algorithms, including the all-pairs algorithm, Johnson's algorithm, the Floyd-Warshall algorithm, A*, the bidirectional Dijkstra and traveling salesperson algorithms, and more.

The PostgREST RESTful API for PostgreSQL databases

PostgREST is a web server created by *Joe Nelson* that effectively turns any PostgreSQL database into an API. This tool manages client authentication through **JSON Web Tokens (JWTs)**. It also securely embeds a database role name in the JWT and uses that database role for all connections. The latest PostgREST release is version 7.0.1, and one of its earliest versions was PostgREST 0.2.6 from February 18, 2015.

PostgREST is built on three core concepts:

- JSON encoding
- Authentication by JWTs
- Resource embedding

Taking advantage of the very fast PostgreSQL JSON encoder for a large number of responses, PostgREST is also a very fast RESTful API. At first, when PostgREST connects to a database, it connects using an authenticator role that only has login capability. Then, whenever an authenticated request is sent to PostgREST that contains a JWT token, the token will be decoded using the secret key to set up a database role for the request. Using the `&select=` parameter, PostgREST can also query the related tables for us through defined foreign keys that have been declared inside the PostgreSQL database.

PostgREST's philosophy is that it aims to do one thing well: add an HTTP interface to any PostgreSQL database.

An overview of PostgreSQL administration

The daily tasks of any PostgreSQL database administrator can include the optimization of parameter values defined by the PostgreSQL architecture, setting up and maintaining PostgreSQL HA clusters, and creating dashboards to monitor their PostgreSQL database clusters. They can also get requests to proceed with performance tests a few times per month if their PostgreSQL databases frequently encounter heavy traffic loads.

The PostgreSQL architecture

The PostgreSQL architecture is very simple. It consists of shared memory and a few background processes and data files, as shown in the following diagram:

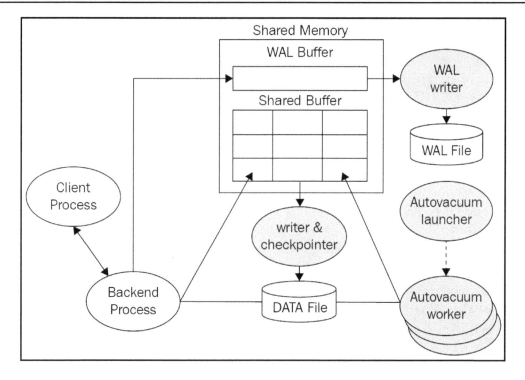

Figure 1.1 – The PostgreSQL architecture

For a better understanding, let's discuss the preceding architecture in the following sections.

Shared memory

Shared memory refers to the memory that is reserved for database caching and transaction log caching. The most important elements in shared memory are shared buffers and WAL buffers:

- **Shared buffers**: The purpose of a shared buffer is to minimize DISK I/O. For this purpose, the following principles must be met:
    - You need to access very large buffers (that is, tens or hundreds of gigabytes worth) quickly.
    - You should minimize contention when many users access it at the same time.
    - Frequently used blocks must remain in the buffer for as long as possible.

- **WAL buffers**: The WAL buffer is a buffer that temporarily stores changes to the database:
  - The contents stored in the WAL buffer are written to the WAL file at a predetermined point in time.
  - From a backup and recovery point of view, WAL buffers and WAL files are very important.

PostgreSQL autovacuum

PostgreSQL includes the VACUUM statement, which reclaims waste storage using dead tuples inside the database. This is because when we delete tuples, they are not physically removed from their table until a vacuum is called. So, on frequently updated tables, we need to perform VACUUM periodically.

The PostgreSQL autovacuum launcher is an optional daemon. By default, it is enabled inside the PostgreSQL configuration using the `autovacuum = on` parameter. Therefore, unless you turn it off, the autovacuum launcher will wake up every `autovacuum_naptime` seconds.

Whenever the autovacuum occurs, by default, every 1 minute, it invokes multiple processes. The number of concurrent vacuum processes is defined by the `autovacuum_worker` parameter value inside the PostgreSQL configuration file. The vacuum work does not lock on any tables, hence it does not interrupt other database tasks.

The PostgreSQL writer and checkpointer

Inside the shared buffer, there are pages that have been modified, but their changes have not been written to disk yet; those pages are called dirty pages. A checkpoint is a specific point in time when all of the dirty pages in the shared buffer are written to disk by the checkpointer. So, if there is a crash, the database recovery will start from the most recent checkpoint where all the data was surely consistent. The PostgreSQL checkpointer flushes all of the dirty pages at certain intervals to create a checkpoint for necessary recovery.

There are two types of writers in PostgreSQL: the WAL writer and the background writer. Between the two checkpoints, the background writer flushes some dirty pages to disk so that there are always enough clean pages to reduce the amount of blocking that occurs during the queries.

PostgreSQL does not write the updates directly to the data files on disk but on commit; these updates are written to the WAL file sequentially, and later, the checkpointer will write all the dirty blocks to each of the respective data files because data file writing cannot be sequential, so users will not have to wait for the delay to locate the data block inside a data file.

Hence, the WAL writer will write our updates from the shared buffer to the WAL file.

PostgreSQL process types

PostgreSQL has four types of processes:

- **The postmaster process**: The postmaster process is the first process to begin when you start PostgreSQL. It performs recovery, initializes shared memory, and runs background processes. It also creates a backend process when there is a connection request from the client process.
- **The background process**: The list of background processes required for PostgreSQL operations can be found in the following table:

Process	Role
Logger	This writes the error message to the log file.
Checkpointer	When a checkpoint occurs, the dirty data inside the shared buffer is written to the file.
Writer	This periodically writes the dirty buffer to a file.
Wal writer	This writes the WAL buffer to the WAL file.
Autovacuum launcher	This launches a new worker process and carries out vacuum operations on bloated tables.
Archiver	When in Archive.log mode, this copies the WAL file to the specified directory.
Stats collector	This collects DBMS usage statistics such as session execution information (`pg_stat_activity`) and table usage statistical information (`pg_stat_all_tables`).

Table 1.1 – The background processes of PostgreSQL

- **The backend process**: The backend process performs the query request of the user process and then transmits the result. Some memory structures are required for query execution; this is called local memory.
- **The client process**: The client process refers to the background process that is assigned for every backend user connection. Usually, the postmaster process will fork a child process that is dedicated to serving a user connection.

Managing HA in PostgreSQL

Managing HA in PostgreSQL is very important to ensure that database clusters maintain exceptional uptime and strong operational performance so that your data is always available to the application.

Master-slave might be the most basic and easiest HA architecture for developers. It is based on one master database with one or more standby servers. These standby databases will remain synchronized (or almost synchronized) with the master, depending on whether the replication is synchronous or asynchronous.

It is important to understand that PostgreSQL does not have a native mechanism to control failovers, that is, when the master fails automatically, the standby server becomes the new master with downtime as close to 0 as possible. To perform this procedure, there are third-party tools such as `repmgr`, `pgpool-II`, or `patroni`, to name a few.

These tools are placed in a layer above PostgreSQL, and they control the health status of the master; when a problem occurs, these tools fire a series of actions to promote the standby server as the new master.

There are several ways to classify a standby database:

- By the nature of the replication:
  - **Physical standbys**: Disk blocks are copied.
  - **Logical standbys**: The streaming of the data changes.
- By the synchronicity of the transactions:
  - **Asynchronous:** There is a possibility of data loss.
  - **Synchronous:** There is no possibility of data loss; the commits in the master wait for the response of the standby.
- By usage:
  - **Hot standbys**: Support read-only connections; the hot standbys are configured for synchronous-commit mode, so their master server must wait for the hot standbys to confirm that they have consolidated the transaction log (when a commit statement is performed on the master, the progress of the commit will only be completed after all of the hot standbys have finished consolidating the transaction log).
  - **Async standbys**: These kinds of servers are configured by asynchronous-commit mode; therefore, the master server will not wait for the async standbys to consolidate the transaction log (when a commit statement is performed on the master, the progress of the commit will not wait for these async standbys to consolidate).

PostgreSQL uses a stream of WAL records to synchronize the standby databases. They can be synchronous or asynchronous, and the entire database server is replicated.

However, a master-slave setup is not enough to effectively ensure HA, as we also need to handle failures. To handle failures, we need to be able to detect them. Once we know there is a failure, for example, errors on the master, or the master is not responding, then we can select a slave and failover mechanism to it with the smallest amount of delay possible. It is important that this process is as efficient as possible, in order to restore full functionality so that the applications can start functioning again. PostgreSQL itself does not include an automatic failover mechanism, so it will require some custom script or third-party tools for this automation.

After a failover happens, the applications need to be notified accordingly so that they can start using the new master. Additionally, we need to evaluate the state of our architecture after a failover because we can run into a situation where we only have the new master running (for instance, we had a master and only one slave before the issue). In that case, we will need to somehow add a slave so as to recreate the master-slave setup we originally had for HA.

Benchmarking PostgreSQL performance

PostgreSQL is the most advanced open source database, hence, PostgreSQL performance should be the first option to evaluate.

Developers benchmark a PostgreSQL database so that they can check the capability and behavior of the database against their application. Based on the benchmarking plan, different hardware can yield different results. It is very important to separate the benchmarked database server from other servers, such as the servers generating the load or the servers collecting performance metrics. As a part of the benchmarking results, developers will obtain application characteristics, such as the following:

- Is the application read/write-intensive?
- How is the read/write split (60:40)?
- How large is the dataset?
- Is the data and structure representative of the actual production database?

Key performance factors in PostgreSQL

A production environment is consolidated with different components, hardware such as CPU/memory, and operating systems. PostgreSQL is installed to communicate with other components of the production environment. Please bear in mind that if the production environment is not properly configured, the overall performance will be degraded.

Some PostgreSQL queries can run faster or slower depending on the configuration that has been set. The goal of database performance benchmarking is to achieve the largest possible throughput. The key performance factors that affect a database are the workload, resources, optimization, and contention. The workload includes batch jobs, dynamic queries for online transactions, and data analytics queries that are used to generate reports. The workload will be different according to the period of each day, week, or month, and it also depends on the applications. The optimization of every database is unique: it can be a database-level configuration or query-level optimization. Contention refers to the conflict condition because two or more components of the workload could attempt to use a single resource at the same time. If contention increases, the throughput will decrease.

Using pgbench for PostgreSQL benchmarking

The `pgbench` application is a standard benchmarking tool of the software provided alongside the official distribution of PostgreSQL. Using `pgbench`, we repeatedly execute given SQL commands that measure the following performance aspects of your PostgreSQL databases:

- Memory performance and disk performance
- Read/write performance
- Connection performance
- Ad hoc queries' performance (that is, special queries for the needs of services or operational information)

Monitoring PostgreSQL databases

There are several key metrics that you will definitely want to keep track of when it comes to database performance, and they are not all database-specific.

You should keep track of shared buffer usage while reading or updating data. In the shared buffer cache, PostgreSQL checks for the execution of a request. It will take data from the disk if the block is not found in the shared buffer cache. Once this is done, the data will be cached in the shared buffer cache of the database.

Implementing effective database monitoring can offer benefits such as increasing performance, increasing the availability of the application, faster database outage detection, and accurately analyzing storage requirements and index performance.

The DevOps environment for PostgreSQL

DevOps is a new and revolutionary way in which to deploy applications and services, including tools for automation, continuous development, integration, and deployment. Usually, frequent delays can occur in Agile deployment sprints when databases are added to the continuous integration and deployment stages, especially PostgreSQL databases. So, DevOps can provide various tools for PostgreSQL database deployment, automation, and scalability.

Among DevOps' tools, Vagrant and VirtualBox are two pieces of software that focus on automation tasks. They allow you to develop and manage virtual machines and environments in a single workflow. They reduce the development environment setup time. Additionally, the benefits of PostgreSQL in a virtual machine are that they are able to standardize the operating system to provide a consistent environment, and they will completely isolate unrelated databases from each other. Using Vagrant and VirtualBox, it only takes a couple of minutes to create a fresh PostgreSQL database server.

Also, Vagrant integrates with other DevOps configuration management tools such as Ansible and Terraform; these are two infrastructure tools that provide script templates to start PostgreSQL database clusters automatically on the cloud. Ansible and Terraform are good ways in which to automate PostgreSQL in order to speed up the development process.

The goal of database automation by DevOps is clear: your PostgreSQL servers will appear the same. Therefore, your PostgreSQL servers will obtain consistency through the quality assurance, staging, and production phases. Puppet is also a well-known DevOps tool that we can use to manage PostgreSQL configuration files, the Postgres user, and PostgreSQL backup crontabs. In comparison, Jenkins, a DevOps tool, can be used to deploy and test virtually any software project. Jenkins can also contribute to the automation of PostgreSQL test databases during software testing tasks.

PostgreSQL testing frameworks

PostgreSQL testing frameworks allow developers to process automated tests for existing stored procedures with unit tests and **Test-Driven Development (TDD)**. Because all test cases are saved in our PostgreSQL databases, they will not be stored by files, version control utilities, or external resources. One test case is a group of many database tests, and each test case is executed independently. After a test case has been executed, the changes that it has made are automatically rolled back.

There are four famous test frameworks for PostgreSQL, as follows:

- `pgTap`: This is a unit test framework for PostgreSQL.
- `PGUnit`: This is a simpler testing tool than `pgTap`.
- Simple `pgunit`: This offers another approach for unit tests.
- `Testgres`: This is a Python testing framework for PostgreSQL.

These testing frameworks allow users to really verify the structure of their database schema, to exercise any views, procedures, functions, rules, or triggers. Furthermore, the `testgres` Python tool can also start a replica for a PostgreSQL database and even execute performance benchmarks for the database.

Summary

In this chapter, we learned about PostgreSQL development. We discovered how PostgreSQL is a popular DBaaS among the clouds and explored the good features of PostgreSQL through various application environments such as NodeJS, Django, PostGIS, and PostgREST. Next, we explored the administration of PostgreSQL by understanding the PostgreSQL architecture, PostgreSQL HA clusters, and PostgreSQL performance. Later, we learned the definition and/or principles of PostgreSQL database monitoring and introduced various DevOps tools for PostgreSQL. Finally, we developed an understanding of popular PostgreSQL database testing frameworks.

In the next chapter, we will learn how to set up a PostgreSQL RDS on a famous cloud – AWS – for ATM machines. This RDS will then be used in the remaining chapters of this book.

Section 2 - Development in PostgreSQL

This section discusses aspects of PostgreSQL development, such as how PostgreSQL can be a popular DBaaS among the current clouds; how to develop PostgreSQL applications in various environments, such as Node.js and Django; how to develop a PostgreSQL database using the PostGIS extension; and how to use the PostgREST standalone web server to turn a PostgreSQL database into a RESTful API.

This section contains the following chapters:

- Chapter 2, *Setting Up a PostgreSQL RDS for ATMs*
- Chapter 3, *Using PostgreSQL and Node.js for Banking Transactions*
- Chapter 4, *Managing Bank ATM Locations Using PostgreSQL and Django*
- Chapter 5, *Creating a Geospatial Database Using PostGIS and PostgreSQL*
- Chapter 6, *Managing Banking Transactions Using PostgREST*

Setting Up a PostgreSQL RDS for ATMs

2

Amazon Web Services (**AWS**) is a famous global cloud computing platform. AWS started delivering web services for IT infrastructures in 2006. Now we call these IT infrastructure services cloud computing. The main benefit of cloud computing is the replacement of upfront infrastructure expenses with low variable costs depending on your own usage. With the cloud, there is no need for us to plan for servers and other IT infrastructures in advance because AWS can spin up and deliver many servers within seconds according to your needs.

There are over 175 IT infrastructure services of AWS, including **Elastic Compute Cloud** (**EC2**), **Elastic Container Service** (**ECS**), **Relational Database Service** (**RDS**), **Simple Storage Service** (**S3**), and CloudFront. Now, AWS supplies reliable and low-cost infrastructure services for thousands of businesses in 190 countries around the world.

In this chapter, we will describe the **Database as a service** (**DBaaS**) options for PostgreSQL through Amazon RDS for PostgreSQL. With Amazon RDS, you can deploy scalable PostgreSQL deployments in just a few minutes with cost-efficient and resizable hardware capacity. Amazon RDS manages complex and time-consuming administrative tasks, such as PostgreSQL software installations and upgrades, storage management, replications for high availability and read throughputs, and backups for disaster recovery. Amazon RDS supports PostgreSQL's major version 12, which includes a number of enhancements to performance, robustness, transaction management, query parallelism, and more.

In this chapter, we will cover the following main topics:

- An overview of the project
- Creating a PostgreSQL RDS with AWS
- Connecting to a PostgreSQL database instance
- Creating a PostgreSQL database snapshot

- Deleting a PostgreSQL database instance
- Restoring data from a PostgreSQL database snapshot
- Point-in-time recovery for PostgreSQL

Technical requirements

The code files for this chapter can be found in the GitHub repository of this book: `https://github.com/PacktPublishing/Developing-Modern-Database-Applications-with-PostgreSQL/tree/master/Chapter02`.

An overview of the project

In this step-by-step project, you will learn how to use PostgreSQL version 12 on the AWS cloud. You will start by creating an empty PostgreSQL database on AWS. Then, you will use the standard pgAdmin software to connect to your AWS database so that you can create a data table and insert all of the ATM data of New York City, which has been provided in our GitHub link. After that, you will discover how to work with this PostgreSQL database by backing up its data into snapshots; hence, you can delete the database and restore it back from your snapshot. Finally, supposing there are some challenging database disasters, you will learn how to perform a point-in-time recovery using AWS.

The project should only take you between 3 and 4 hours to complete. However, if this is the first time you are working with AWS, you might prefer to repeat the steps a few times so that you can carefully study DBaaS in the AWS cloud.

We are going to set up a PostgreSQL 12 RDS for the storage of bank ATM machine locations inside the city of New York. There are around 654 bank-owned ATM locations in New York City – part of the data can be seen in *Table 2.1*. The source of our ATM data is granted by the Open Data NY program, which gives access to government data and information. We have already received permission from the Open Data NY public service to use part of their data for this book. From the Open Data NY public data, we use the following web link to filter all of the ATM locations of New York City: `https://data.ny.gov/Government-Finance/Bank-Owned-ATM-Locations-in-New-York-State/ndex-ad5r/data`.

We will need to import the following table contents into our PostgreSQL database:

ID	Name of the institution	Street address	County	City	State	ZIP code
1	Wells Fargo ATM	500 W 30 STREET	New York	New York	NY	1000
2	JPMorgan Chase Bank, National Association	1260 Broadway	New York	New York	NY	10001
3	Sterling National Bank of New York	1261 Fifth Avenue	New York	New York	NY	10001
4	Bank of America N.A. GA1-006-15-40	1293 Broadway	New York	New York	NY	10001
5	Bank of Hope	16 West 32nd Street	New York	New York	NY	10001
6	TD Bank N.A.	200 West 26th Street	New York	New York	NY	10001
7	Citibank N. A.	201 West 34th Street	New York	New York	NY	10001
8	Capital One, N.A.	215 West 34th Street	New York	New York	NY	10001
9	Citibank N. A.	22 West 32nd Street	New York	New York	NY	10001
10	Sterling National Bank of New York	227 West 27th Street	New York	New York	NY	10001
11	JPMorgan Chase Bank, National Association	245 Seventh Avenue	New York	New York	NY	10001
12	Amalgamated Bank	275 Seventh Avenue	New York	New York	NY	10001
13	JPMorgan Chase Bank, National Association	305 Seventh Avenue	New York	New York	NY	10001
14	Woori America Bank	330 Fifth Avenue	New York	New York	NY	10001
15	Commerce Bank, N.A.	341 Ninth Avenue	New York	New York	NY	10001
16	TD Bank N.A.	350 West 31st Street	New York	New York	NY	10001
17	USAlliance Financial	350 West 31st Street	New York	New York	NY	10001
18	Bank of America N.A. GA1-006-15-40	358 Fifth Avenue	New York	New York	NY	10001
19	Sterling National Bank of New York	406 West 31st Street	New York	New York	NY	10001
20	Sterling National Bank of New York	7 Penn Plaza	New York	New York	NY	10001
...	...	...	...	...	...	...
635	TD Bank N.A.	1709 Third Avenue	New York	New York	NY	10128
636	Citibank N. A.	1781 First Avenue	New York	New York	NY	10128
637	JPMorgan Chase Bank, National Association	1801 Second Avenue	New York	New York	NY	10128
638	JPMorgan Chase Bank, National Association	181 East 90th Street	New York	New York	NY	10128
639	Citibank N. A.	340 East 93rd Street	New York	New York	NY	10128
640	HSBC Bank USA, National Association	45 East 89th Street	New York	New York	NY	10128
641	Bank of America N.A. GA1-006-15-40	345 Park Avenue	New York	New York	NY	10154
642	HSBC Bank USA, National Association	617 Third Avenue	New York	New York	NY	10158
643	Citibank N. A.	200 Park Avenue	New York	New York	NY	10166
644	Bank of America N.A. GA1-006-15-40	200 Park Avenue	New York	New York	NY	10166
645	Apple Bank For Savings	122 East 42nd Street	New York	New York	NY	10168
646	JPMorgan Chase Bank, National Association	405 Lexington Avenue	New York	New York	NY	10174
647	Bank of America N.A. GA1-006-15-40	425 Lexington Avenue	New York	New York	NY	10174
648	HSBC Bank USA, National Association	101 Park Avenue	New York	New York	NY	10178
649	HSBC Bank USA, National Association	117 Broadway	New York	New York	NY	10271
650	Valley National Bank	120 Broadway	New York	New York	NY	10271
651	Citibank N. A.	120 Broadway	New York	New York	NY	10271
652	JPMorgan Chase Bank, National Association	331-337 South End Avenue	New York	New York	NY	10280
653	USAlliance Financial	200 Vesey Street	New York	New York	NY	10281
654	TD Bank N.A.	90 Fifth Avenue	New York	New York	NY	11011

Table 2.1: Bank ATM locations in New York City

We will now move on to the step-by-step project of PostgreSQL creation with AWS.

Creating a PostgreSQL RDS with AWS

Once you register yourself as a user with AWS, sign in and navigate to the AWS Management Console using the following link: `https://console.aws.amazon.com/` `console`.

Amazon RDS stands for **Amazon Relational Database Service**. RDS provides an easy way to set up, operate and/or scale a relational database in the cloud.

Creating a PostgreSQL database

By going through the following steps in detail, you will be able to create a new PostgreSQL RDS and grant access to your PostgreSQL RDS:

1. Click on the **Services** drop-down menu option and navigate to the **Database** section, as shown in the following screenshot:

Figure 2.1 – AWS Services

2. Then, click on the first entry of the **Database** section named **RDS**.

3. We are proceeding to create a PostgreSQL version 12 RDS; you can scroll through the different zones of AWS to select any cloud zones, as shown in the following screenshot. Here, we will keep the zone as it is, that is, **N. Virginia**:

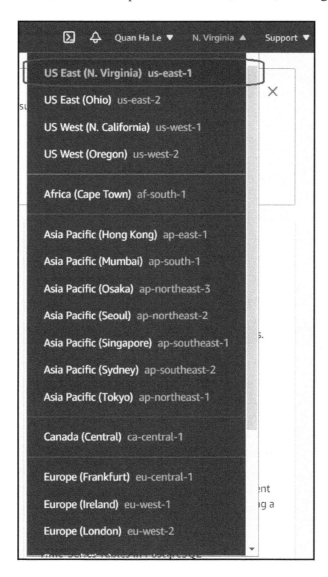

Figure 2.2 – AWS cloud zones

4. You can either click on the **Databases** tab on the left-hand panel or you can scroll down the page and click on the **Create database** button, as shown in the following screenshot:

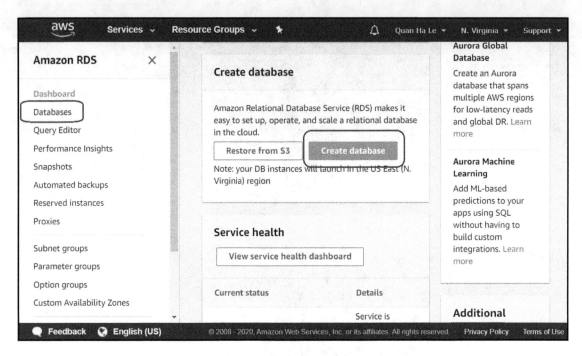

Figure 2.3 – RDS Services

5. If you clicked on the **Create database** button in *step 4*, skip this step. Otherwise, if you selected the **Databases** tab on the left-hand side, you will now reach the following **Databases** page:

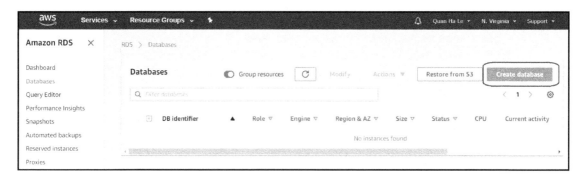

Figure 2.4 – The Databases tab

Click on the **Create database** button on the right-hand side.

6. The next step is to select the **PostgreSQL** database engine, as shown in the
    following screenshot:

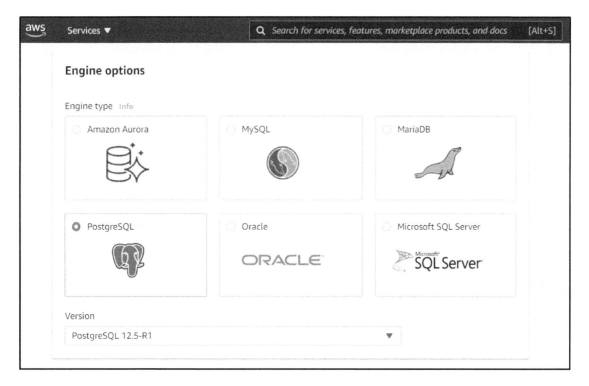

Figure 2.5 – The PostgreSQL version 12 database engine

7. Keeping the **Version** as **PostgreSQL 12.5** and select Free tier template for your database as shown in the following screenshot:

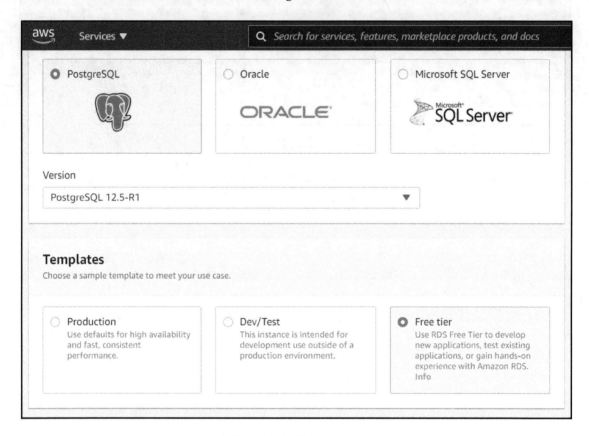

Figure 2.6 – PostgreSQL database details

8. Now, scroll down and continue to set the following values:

- **DB instance identifier**: atm
- **Master username**: dba
- **Master password**: bookdemo
- **Confirm password**: bookdemo

You can view these options in *Figure 2.7*:

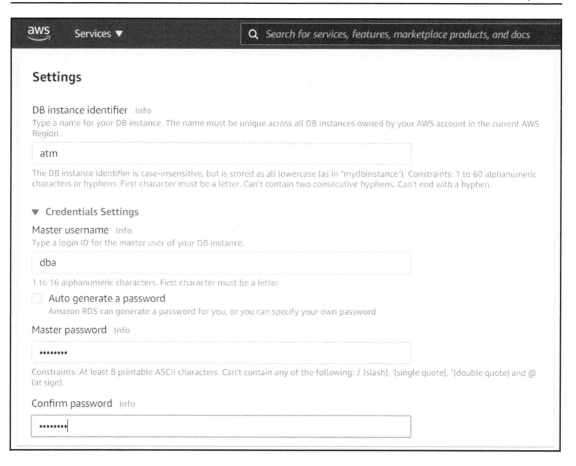

Figure 2.7 – PostgreSQL database details (cont.)

9. After you scroll down, in the **DB instance class**, select **Burstable classes** and enter the following values:

- **DB instance class**: db.t2.micro
- **Storage type**: General Purpose (SSD)
- **Allocated storage**: 20 (GiB)

This is shown in *Figure 2.8*:

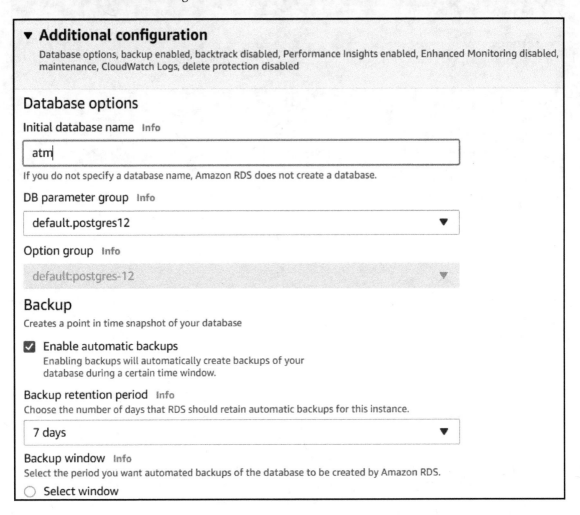

Figure 2.8 – PostgreSQL database details (cont.)

10. Expand the **Additional connectivity configuration** section and set the value to **Yes** for **Publicly accessible**.

11. Expand the **Additional configuration** section and enter `atm` for **Initial database name**. For **Backup retention period**, the default value of **7 days** means that Amazon RDS should retain automatic snapshot backups of this PostgreSQL database for **7 days**. You can simply leave it to **7 days**, or you can change the value to a longer or shorter period. If you select **0 days**, this means that you will not be utilizing the automatic RDS backups:

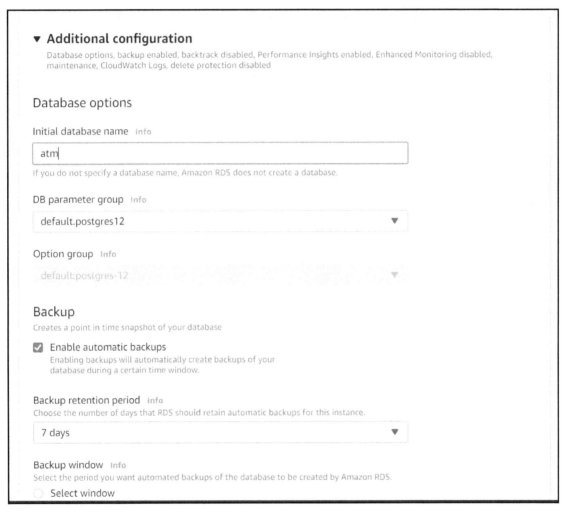

Figure 2.9 – PostgreSQL additional configuration

12. For the remaining options, you can use the default values of AWS. Afterward, scroll down and click on the **Create database** button:

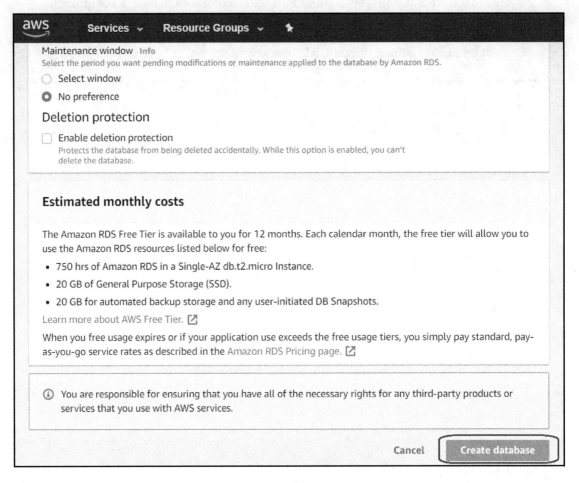

Figure 2.10 – Creating a PostgreSQL database instance

13. Wait until the status of the database instance changes from **Creating** to **Available**, as shown in the following screenshot (this may take a few minutes):

Figure 2.11 – Creating a PostgreSQL RDS

14. When the database status changes to **Available**, click on the **atm** database to note down the RDS endpoint, as shown in the following screenshot:

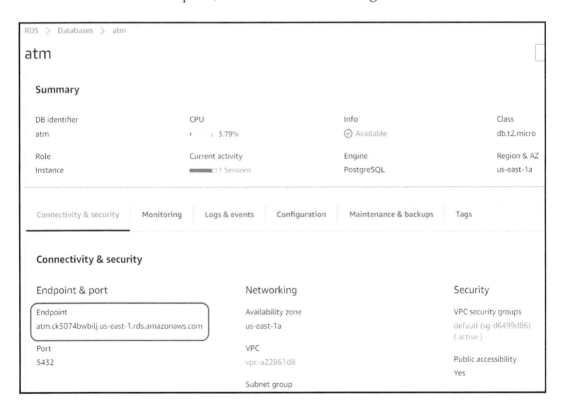

Figure 2.12 – The PostgreSQL RDS endpoint

As with our example, the endpoint is **atm.ck5074bwbilj.us-east-1.rds.amazonaws.com**.

After the RDS has been created, we will not be able to use the PostgreSQL database right away because we have not had the security permissions to access the RDS yet. In the next section, we will edit the security group of this new RDS to add our access permit.

Editing the RDS security group

In order to allow all of your work locations to access your new RDS, you have to add all of their IP addresses to the RDS security group:

1. Recheck the database's security group to make sure that you are able to connect to the RDS from your working locations. Click on **VPC security groups default (sg-d6499d86)**, as shown in the preceding screenshot. This will redirect you to the security group page where you can click on the **Inbound** tab, as shown in the following screenshot:

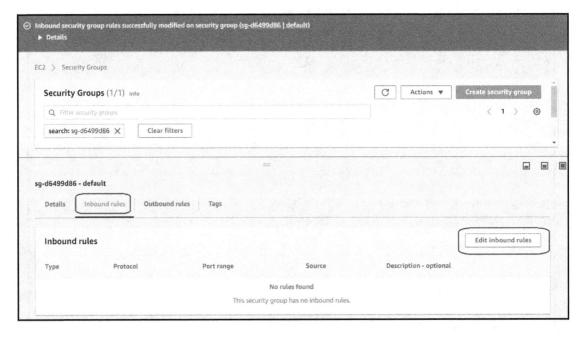

Figure 2.13 – The PostgreSQL RDS security group

2. Now click on the **Edit inbound rules** button, as shown in the preceding screenshot. Add a rule to add your current IPs by entering the following values and clicking on **Save rules**:

- **Type**: `PostgreSQL`
- **Protocol**: `TCP`
- **Port Range**: `5432`
- **Source**: `MyIP` or `Custom`
- **IP**: your other IP location

This is shown in the following screenshot:

Figure 2.14 – The RDS security group's rules

By default, AWS allows your current IP address to access your RDS endpoint; in our case, our current IP is **70.77.107.46**. If you have more than one work location, you can add more IP addresses from all of your work locations, as shown in the preceding screenshot.

In this section, we have demonstrated how to, step by step, create a standalone PostgreSQL 12 database on the well-known AWS cloud and then grant access to the RDS security group. Now that you have the correct access to the RDS security group, in the next section, we will show you how to use other PostgreSQL tools such as pgAdmin to connect to your new PostgreSQL RDS.

Connecting to a PostgreSQL database instance

Here, we are using a PostgreSQL tool called pgAdmin 4, version 4.5, which was released in April 2019, to insert data for our ATM database. Download and install the pgAdmin 4 tool from `https://www.pgadmin.org/download/pgadmin-4-windows/`, and you should be able to view the main browser of pgAdmin 4.5, as follows:

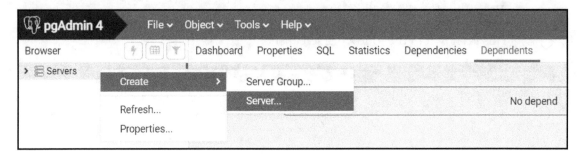

Figure 2.15 – The pgAdmin 4.5 main browser

Next, we need to create a connection to the RDS before we can read and write data to the created ATM database.

The pgAdmin server

We can connect the RDS to pgAdmin by performing the following steps:

1. From the pgAdmin 4.5 main browser, right-click on **Servers**. Then, select **Create** and click on **Server...**; the **Create - Server** popup will display, as follows:

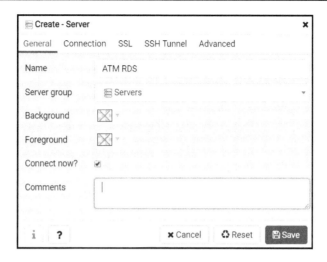

Figure 2.16 – The Create - Server popup

2. Type the name ATM RDS into the **Create - Server** popup. Then, select
   the **Connection** tab, as shown in *Figure 2.17*:

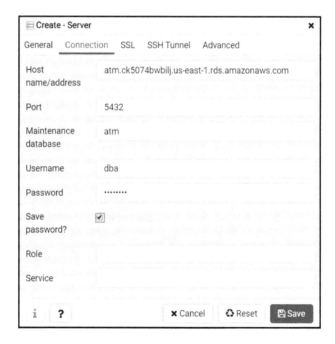

Figure 2.17 – PostgreSQL Connection

3. Within the **Connection** tab shown in the preceding screenshot, enter your PostgreSQL RDS details:

- **Host name/address** (your RDS endpoint): `atm.ck5074bwbilj.us-east-1.rds.amazonaws.com`
- **Username** (your RDS master username): `dba`
- **Password** (your RDS password): `bookdemo`

   Once these details have been entered, click on **Save**.

4. Expand the servers in your pgAdmin by clicking on the **>** button on the left-hand side, and then you can connect to the ATM RDS by simply double-clicking on the **ATM RDS** link. Alternatively, you can right-click on it and then click on **Connect Server**:

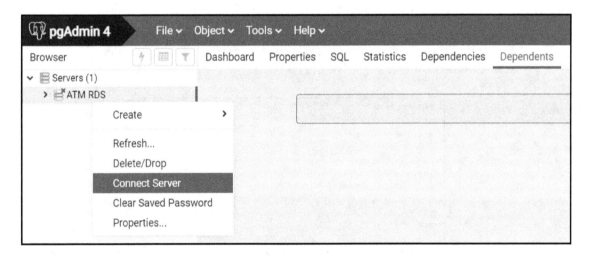

Figure 2.18 – Connecting from pgAdmin to the PostgreSQL RDS

5. Expand the **Databases** drop-down menu inside your pgAdmin, and then select the **atm** database. After that, continue to navigate to **atm** > **Schemas** > **public**, as shown in the following screenshot:

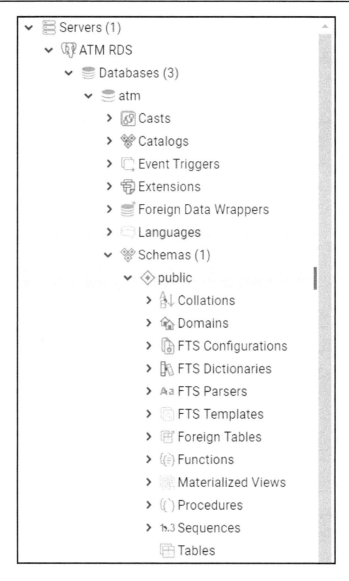

Figure 2.19 – The PostgreSQL schemas on pgAdmin

Now that we have used pgAdmin to connect to the RDS from AWS, we will move on to the next section, where we will create the ATM locations table and populate it with ATM data.

The ATM locations table

The pgAdmin tool offers a graphical user interface to create a table, so we are going to use that user interface instead of DDL statements:

1. We will create the ATM locations table inside the ATM RDS. For this, we need to right-click on the **Tables** entry on the left-hand side. Then, click on **Create** and select **Table**..., as shown in the following screenshot:

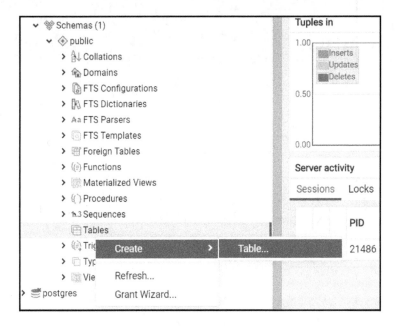

Figure 2.20 – Creating a table with pgAdmin

2. Within the following **Create - Table** dialog box, enter ATM locations inside the **Name** section:

Figure 2.21 – The Create - Table dialog box

3. Select the **Columns** tab, and then use the add new row (*Shift* + *Ctrl* + *A*) button (**+**) on the right-hand side to add all the table columns, as shown in the following screenshot. Then, click on **Save**:

Figure 2.22 – Creating columns for the ATM locations table

4. After the ATM locations table has been created, select the **Tools** menu and then select **Query Tool**, as shown in the following screenshot:

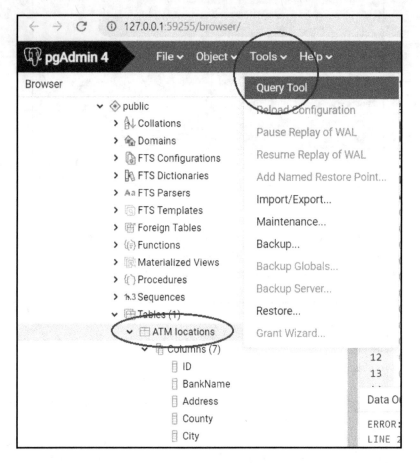

Figure 2.23 – Using Query Tool

5. Now, copy the content of the `atmlocations.sql` SQL script from the GitHub link
   at `https://github.com/lequanha/PostgreSQL-12-Development-and-Administra`
   `tion-Projects/blob/master/Chapter 2/atmlocations.sql`.

Next we paste the `atmlocations.sql` file content into the Query Tool and click the Execute/Refresh (F5) ▶ icon to insert all the data in as shown in the following screenshot:

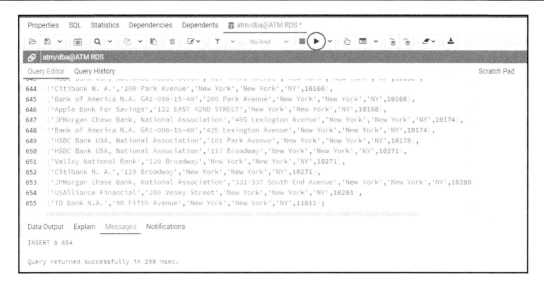

Figure 2.24 – Inserting data into the ATM locations table

6. Click on the data view icon, ⊞ , to display the whole table, as shown in the following screenshot:

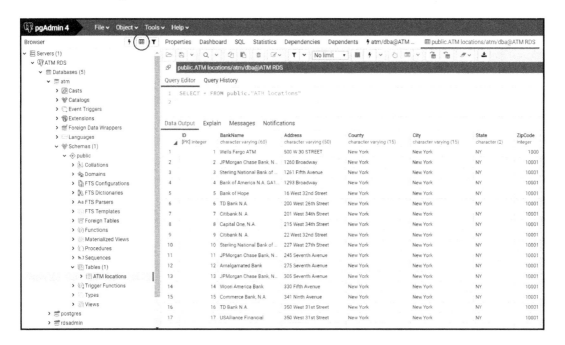

Figure 2.25 – The data of the ATM locations table

In this section, we have guided you through how to populate the AWS PostgreSQL 12 RDS by using pgAdmin; this step-by-step guide is simple and the SQL script is supplied by GitHub. In order to keep your newly added data safe from incidents, you will need to back up your database. The AWS cloud offers a backup facility called a snapshot so that you can back up your RDS. In the next section, we will show you how to create a PostgreSQL snapshot.

Creating a PostgreSQL database snapshot

Backing up a database is a daily task that protects you from possible data disasters. AWS supplies a snapshot method so that you can back up your data not only every day but also every hour or whenever you need to implement a database backup. Perform the following steps to back up your RDS:

1. From the **Amazon RDS Databases** environment, select the **atm** database and then select the **Actions** option. Then, click on **Take snapshot**, as shown in the following screenshot:

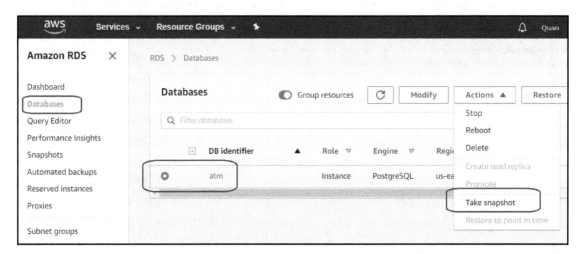

Figure 2.26 – Taking an RDS snapshot

2. Enter a valid name for the intended snapshot and then click on the **Take Snapshot** button:

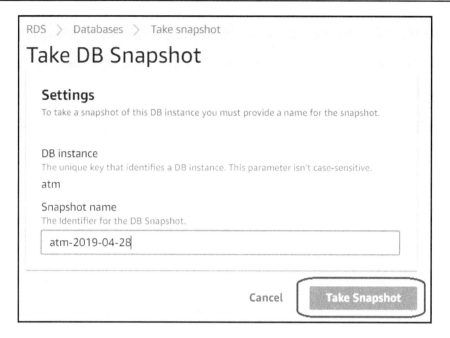

Figure 2.27 – Adding a snapshot name

3. Wait for a short while until the snapshot status on the next web page changes from **Creating** to **available**, as shown in the following screenshot:

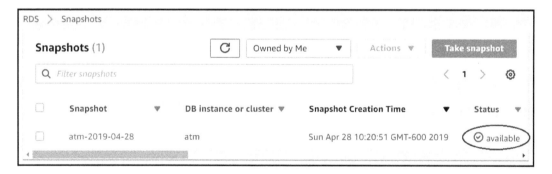

Figure 2.28 – Snapshot status

The preceding snapshot outcome can be used to restore the whole RDS, including all of the current structures and data that we have created and populated so far. However, before we can practice snapshot restoration, we need to delete our RDS. Then, we can illustrate how snapshot restoration recovers the same database with your same data. Therefore, in the next section, we will demonstrate the next step of how to delete a PostgreSQL RDS.

Deleting a PostgreSQL database instance

Sometimes, you might want to remove a piece of software and, in doing so, also remove the unused database from your AWS account. You can delete a PostgreSQL RDS that does not have deletion protection enabled. When we created or modified our PostgreSQL RDS in the previous sections, there was an option to enable deletion protection so that the RDS cannot be deleted. Perform the following steps:

1. Click on **Databases** from the navigation tabs on the left-hand side. Then, select the **atm** database to enable the **Actions** menu, as shown in the following screenshot:

Figure 2.29 – Enabling the Actions menu on the atm database

2. Check the drop-down menu for **Actions** and select the **Delete** option, as shown in the following screenshot:

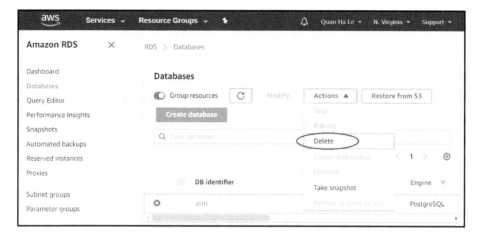

Figure 2.30 – The Actions > Delete option to remove the atm database

3. Enter delete me to confirm the deletion, as shown in the following screenshot. Then, click on **Delete**:

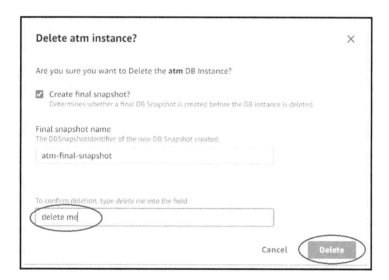

Figure 2.31 – Delete confirmation

You can choose to create one more final snapshot before deletion, or if you already have a good snapshot just as we created manually in the *Creating a PostgreSQL database snapshot* section, you will be able to uncheck the final snapshot creation.

4. Click on **Snapshots** from the left-hand side navigation tabs when the **atm** database has been deleted in order to view all the snapshots that you have so far, as shown in the following screenshot:

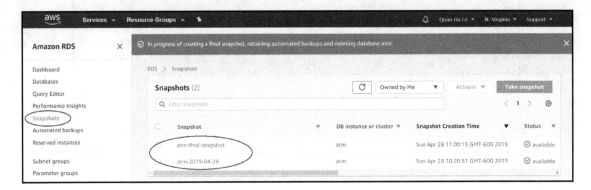

Figure 2.32 – A list of snapshots

In this section, we showed you how to safely delete the PostgreSQL 12 RDS because we have stored all of the previous databases within good snapshots already. In practice, when your database loses data or it crashes, you will rely on your snapshots to recover your good RDS PostgreSQL database again. In the next section, we will show you how to restore your RDS from a PostgreSQL snapshot.

Restoring data from a PostgreSQL database snapshot

In the case of database incidents leading to data loss, you can use your RDS snapshot to restore a new PostgreSQL RDS. The recovered RDS will contain the same ATM database with the exact same tables and data rows that your former RDS had when you created the snapshot. The username and password of your RDS automatically remain the same. However, the default security group is applied to your restored RDS; therefore, you will have to modify or correct the security group to grant access to your connections. Perform the following steps for snapshot restoration:

1. Starting from *Figure 2.32*, select the snapshot that you would like to use to restore the deleted **atm** database, for example, the **atm-2019-04-28** snapshot.
2. Click on the drop-down menu named **Actions** and select **Restore Snapshot**:

Figure 2.33 – Database restoration

3. Under the **Settings** and **DB instance size** sections of the **Restore DB Instance** page that is shown, enter the following database details:

- **DB instance identifier**: atm
- **DB instance class**: db.t3.micro

This is shown in the following screenshot:

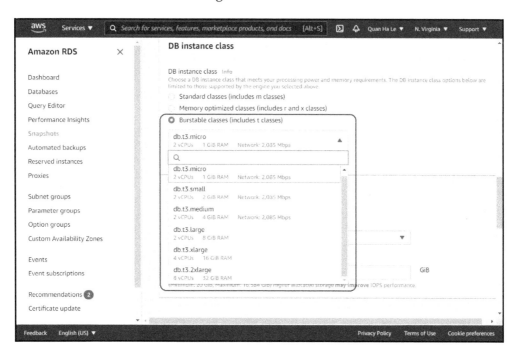

Figure 2.34 – The DB instance size section of the database restoration page

Now, scroll down to the bottom of the page and click on the **Restore DB Instance** button, as shown in the following screenshot:

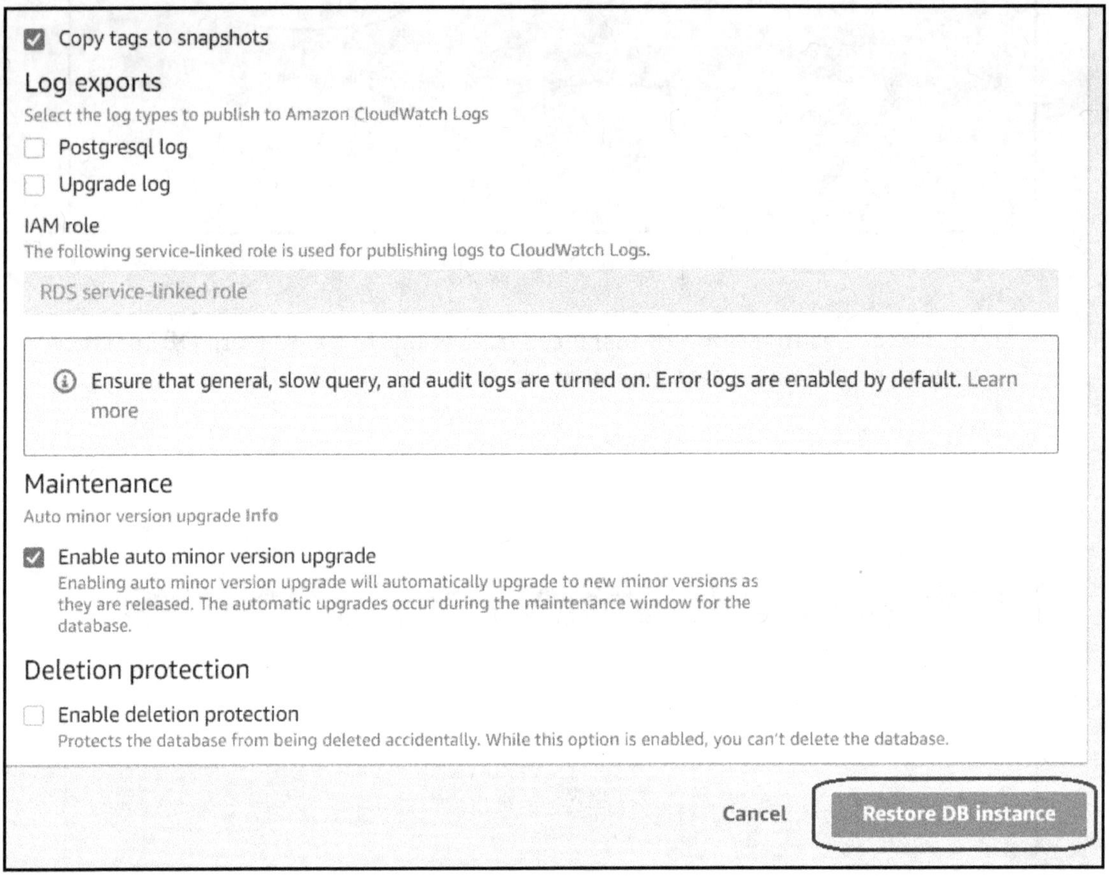

☑ Copy tags to snapshots

Log exports

Select the log types to publish to Amazon CloudWatch Logs

☐ Postgresql log

☐ Upgrade log

IAM role

The following service-linked role is used for publishing logs to CloudWatch Logs.

RDS service-linked role

ⓘ Ensure that general, slow query, and audit logs are turned on. Error logs are enabled by default. Learn more

Maintenance

Auto minor version upgrade Info

☑ Enable auto minor version upgrade
Enabling auto minor version upgrade will automatically upgrade to new minor versions as they are released. The automatic upgrades occur during the maintenance window for the database.

Deletion protection

☐ Enable deletion protection
Protects the database from being deleted accidentally. While this option is enabled, you can't delete the database.

Cancel   **Restore DB instance**

Figure 2.35 – Restoring the database instance

4. Wait for a short while until the ATM database is fully available on the next screen.

5. Select the ATM database and click on the **Modify** button to correct the security group for the ATM database by using the **sg-d6499d86** security group that we set up last time for the ATM RDS.

6. Additionally, click on **X** to remove the incorrect default security group and then click on **Continue**:

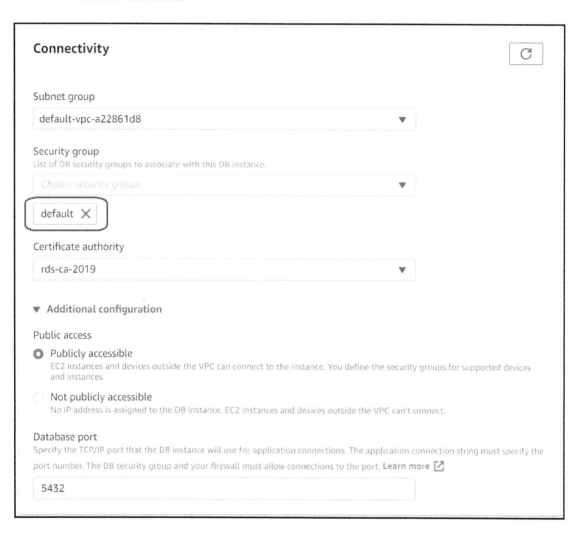

Figure 2.36 – Security group restoration

7. On the next summary page, select the **Apply immediately** option and then click on **Modify DB Instance**, as shown in the following screenshot:

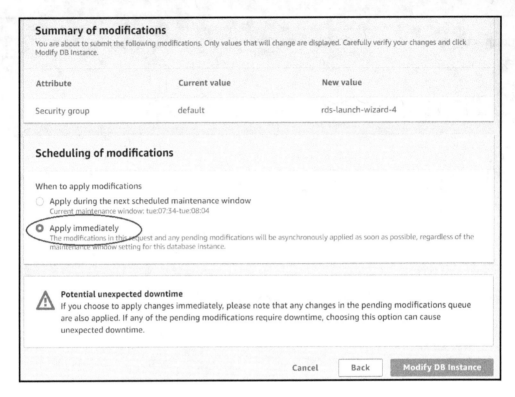

Figure 2.37 – The summary page of RDS modification

8. Wait for a short while until the ATM database's status changes from **Modifying** to **Available** on the next screen.

The username and the password of the RDS that we set up last time are automatically restored:

- **Username** (your RDS master username): dba
- **Password** (your RDS password): bookdemo

Now you can try opening pgAdmin to reconnect to the **atm** database as usual.

In this section, we demonstrated quite an important task inside the daily activities of a PostgreSQL database administrator; hence, this step-by-step guideline is useful for us.

Point-in-time recovery for PostgreSQL

A PostgreSQL RDS can be restored to a specific point in time. The default security group is applied to the newly restored database instance. You will have to modify or correct the security group to grant access to your connections. We can do this by performing the following steps:

1. From the Amazon RDS console in the left-hand navigation pane, choose **Databases**.
2. Select the **atm** database.
3. If you have not done so yet, you can enable automated backups by clicking on the **Modify** button and then changing the **Backup retention period** option from **0 days** to another value from **1 day** to **35 days**, as shown in the following screenshot:

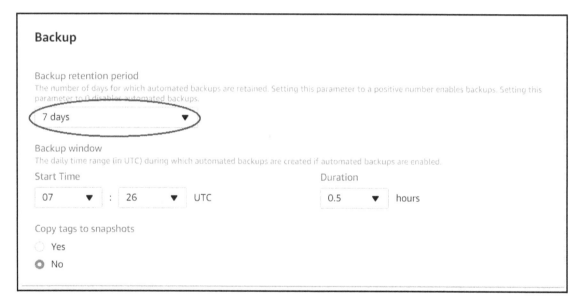

Figure 2.38 – Enabling backup retention

After that, apply the change immediately as you get used to this RDS modification.

4. Once the **atm** database is available, from **Actions**, select **Restore to point in time**, as shown in the following screenshot:

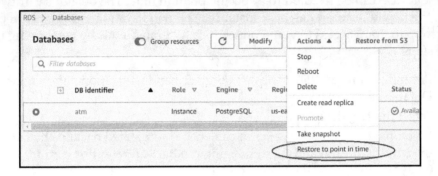

Figure 2.39 – Restore to point in time

5. From the **Launch DB Instance** page, select either of the following options:

- Select **Latest restorable time** to restore to the latest possible time.
- Select **Custom** to choose a specific time, as shown in the following screenshot:

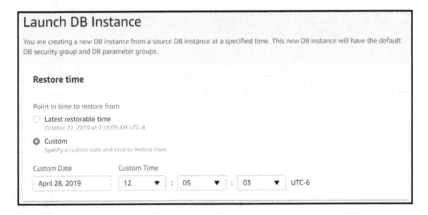

Figure 2.40 – Launch DB Instance

Now, scroll down to the **Instance specifications** section and the **Settings** section on the same page to enter the other options, as follows:

- **DB instance class**: db.t3.micro
- **DB instance identifier**: atmPITR

This is shown in the following screenshot:

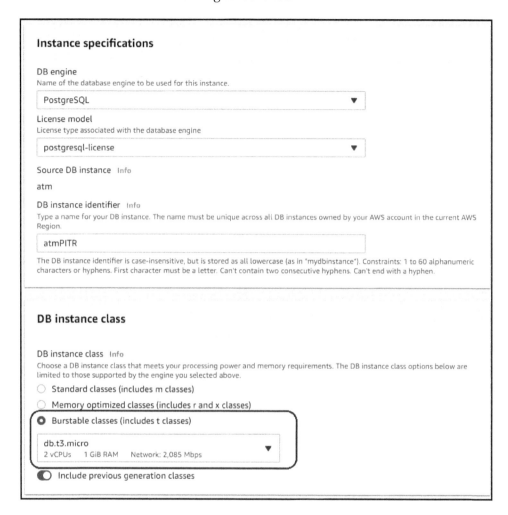

Figure 2.41 – Recovering the database instance settings

After that, navigate to the very end of the page to click on the **Launch DB Instance** button.

6. Wait for a couple of minutes until the status of the newly launched **atmpitr** RDS changes to **Available**, as shown in the following screenshot:

Figure 2.42 – The new point-in-time RDS

In order to grant connection access to the new point-in-time RDS, correct the security group in the same way that you did in the previous section from *steps 5 to 8*.

Point-in-time recovery is especially useful when the database schema structures are modified from time to time. It is not recommended that database schema structures are changed after the applications have already been deployed because the change can cause conflicts with other integrated modules. However, changes in schema structure are sometimes unavoidable. Therefore, after you have changed the schema structures and other application modules are still coded on the ex-structures, the related incompatibilities can pop up improperly. The real difficulty arises if the pop-up issues do not appear right away and you only detect them a few weeks later. In this case, you will be able to restore your RDS back to specific timing points to troubleshoot the related incompatibilities.

Summary

In this chapter, you learned about DBaaS options through Amazon RDS for PostgreSQL. Through a step-by-step project for the storage of banking ATM machine locations inside New York City, you researched how to use Amazon RDS for PostgreSQL DBaaS. Now you are well versed regarding DBaaS within the common tasks of the AWS cloud. You practiced how to create a PostgreSQL RDS and how to connect to that RDS from pgAdmin. Then, you also acquired skills to delete, back up, restore, and maintain a PostgreSQL RDS. You also learned how to restore a database and recover PostgreSQL to a specific point in time.

In the next chapter, we will create a step-by-step NodeJS project and use the PostgreSQL RDS created in this chapter to implement a RESTful API.

3
Using PostgreSQL and Node.js for Banking Transactions

Node.js, Express, and PostgreSQL is a powerful stack for the backend to offer the four basic operations of persistent storage: **create**, **read, update, and delete** (**CRUD**). In this chapter, we will describe the steps to create a RESTful web service that will be an **application programming interface** (**API**) based on Node.js, Express, and PostgreSQL and it will implement HTTP requests such as the GET, PUT, POST, and DELETE methods on data.

We will not only learn how to apply Node.js to create a JavaScript backend, but we will also develop the frontend application using JavaScript in the **Model View Controller** (**MVC**) approach by Angular. Angular is available worldwide and open source and it helps to extend HTML's syntax to clearly express application components. Being cross-browser compliant, Angular suitably handles JavaScript code for different browsers. Therefore, we consider Angular an ideal framework to illustrate our ATM web application.

PostgreSQL uses a controlled set of **access control lists** (or **ACLs**) to decide which users are allowed to select from, to update objects (tables, views, and sequences) within a database. PostgreSQL ACLs are defined by superusers and owners of database objects through the GRANT and REVOKE commands. There are four basic PostgreSQL ACL privileges: SELECT, INSERT, UPDATE, and DELETE. Also, a superuser or a database owner can use ALL to grant all ACL privileges for another user.

Our Node.js and Angular application will directly use the master username dba of the PostgreSQL RDS on AWS to make connections. Because the master user dba is the owner of the atm database inside our PostgreSQL, our Node.js and Angular app will have all of the privileges on the atm database: SELECT, INSERT, UPDATE, DELETE, TRUNCATE, REFERENCES, TRIGGER, CREATE, CONNECT, TEMPORARY, EXECUTE, and USAGE.

The chapter will cover the following topics:

- Setting up a Node.js project
- Setting up PostgreSQL for Node.js
- Working with the server side: routes
- Working with the client side: Angular
- Create an Angular view
- Angular app refactoring
- Automated testing with Mocha and Chai

Technical requirements

This chapter will take developers around 10-12 hours of work to develop an ATM web application.

The completed source code and figures for this chapter can be found on GitHub here: `https://github.com/PacktPublishing/Developing-Modern-Database-Applications-with-PostgreSQL/tree/master/Chapter03`.

Setting up a Node.js project

In Chapter 2, we used **Amazon Web Services** (**AWS**) to set up a PostgreSQL RDS. In this chapter, we will use an AWS EC2 instance to create an API from Node.js to PostgreSQL.

An EC2 instance is a virtual server in **Elastic Compute Cloud** (or **EC2**) for executing our applications on the cloud infrastructure. Hence EC2 is a service evolving cloud computing platform and there are a variety of types of EC2 instances with different configurations of CPU, memory, storage, and networking resources to suit user needs. Each type is also available in two different sizes to address workload requirements.

In order to launch our EC2 instance for Node.js, we need to select an **Amazon Machine Image** (**AMI**), which is a special type of virtual appliance to deploy for services delivered to our EC2 instance. For our Node.js project in this chapter, we will use a good AWS AMI: `ami-7d579d00`, as seen in *Figure 3.1.*

We decide to use a CentOS image because CentOS is a very stable Linux version, its software does not need to be updated too often, and CentOS will guarantee less downtime for our EC2 instance. The preceding image is a free tier AMI using the HVM virtualization type, which provides fast access to the bare-metal hardware platform:

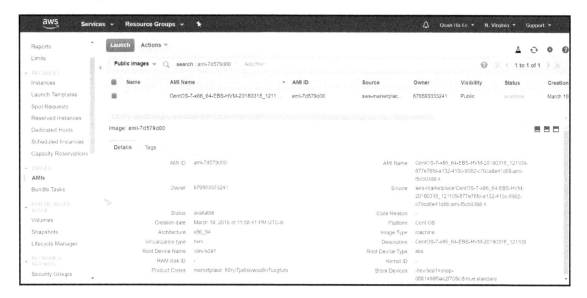

Figure 3.1 – The AMI in usage

The EC2 instance for Node.js can be launched in the same – or not in the same – AWS zone from the PostgreSQL 12 RDS in the Database Preview Environment. You can launch it from the Ohio zone or N. Virginia zone – whichever best suits your needs.

We PuTTY into our EC2 instance with the built-in user for this AMI: `centos user`. `centos user` is a built-in account of this AMI. Then, we are going to switch to the user root:

```
[centos@ip-172-31-95-213 ~]$ sudo su
[root@ip-172-31-95-213 centos]# cd /usr/local/src/
```

Now we are in the `src` folder to implement our expected projects for all chapters and we are under the root user so that we can install all necessary packages smoothly.

Installing Node.js

We'll install Node.js using the following steps:

1. To install Node.js, add a Node.js yum repository for the stable version 10.x:

```
[root@ip-172-31-95-213 src]# yum install wget
Please enter 'y' for yes when asked
[root@ip-172-31-95-213 src]# yum install -y gcc-c++ make
[root@ip-172-31-95-213 src]# curl -sL
https://rpm.nodesource.com/setup_10.x | sudo -E bash -
```

The Yum repository is provided by the Node.js official website. After adding the Yum repository, we can start installing Node.js.

2. Install Node.js on CentOS with yum:

```
[root@ip-172-31-95-213 src]# yum install nodejs
Please enter 'y' for yes when asked
```

When Node.js is installed, NPM will be included as well as many other dependent packages on your system.

3. Check the Node.js and NPM version to make sure that our installation has been successful:

```
[root@ip-172-31-95-213 src]# node -v
v10.15.3
[root@ip-172-31-95-213 src]# npm -v
6.4.1
[root@ip-172-31-95-213 src]# mkdir node-api-postgres
[root@ip-172-31-95-213 src]# cd node-api-postgres
[root@ip-172-31-95-213 node-api-postgres]#
```

4. Run npm init -y to create a package.json file as in *Figure* 3.2. A package.json file in your package helps make it easy for other developers to manage and install:

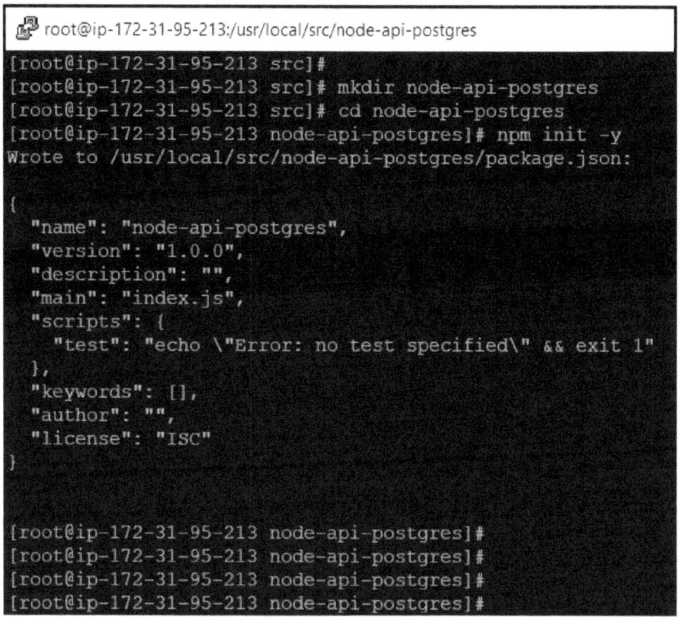

Figure 3.2 – Creating a package.json file

The `package.json` file usually lists the packages your project depends on, specifies versions of a package that your project can use, and it makes your build reproducible, and therefore your project will be easier to share with other developers.

RESTful web services

A RESTful API is an API that uses HTTP requests to GET, PUT, POST, and DELETE data. Although SOAP has been the preferred and chosen option by many companies, for others it is too complex and inflexible. For this reason, REST-based services are beginning to be used to display massive amounts of data and this is the goal that we have in this book.

Our design of a RESTful API for Node.js will include the endpoints shown in *Table 3-1*.

URL	HTTP Verb	Action
/api/atm-locations	GET	Return ALL ATM machines
/api/atm-locations/:id	GET	Return a SINGLE ATM machine
/api/atm-locations	POST	Add an ATM machine
/api/atm-locations/:id	PUT	Update an ATM machine
/api/atm-locations/:id	DELETE	Delete an ATM machine

Table 3.1 – The RESTful services

Let's check it out using the following steps:

1. Let's create an Express Node.js app with the help of the following command:

```
[root@ip-172-31-95-213 node-api-postgres]# npm install express-
generator -g
```

2. Express is a minimal and flexible Node.js web application framework that facilitates the rapid development of Node-based web applications. Now install Express in the `node-api` directory and save it in the dependencies list:

```
[root@ip-172-31-95-213 node-api-postgres]# npm install express -g
[root@ip-172-31-95-213 node-api-postgres]# express node-api
```

The Express framework will now quickly generate the whole tree of our API project, as shown in *Figure 3.3*:

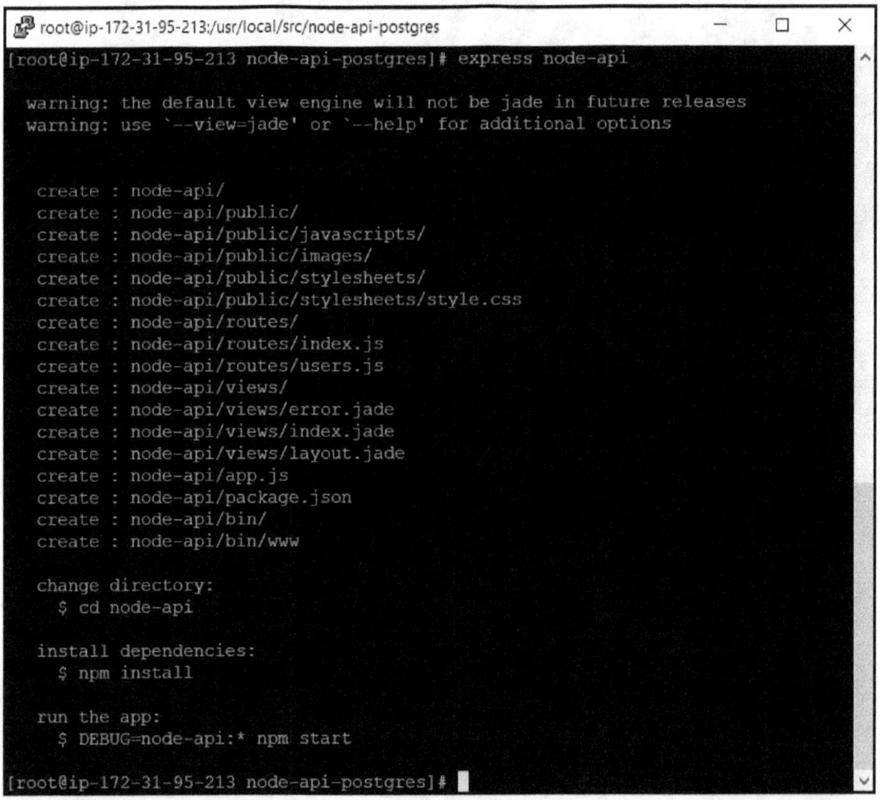

Figure 3.3 – Creating an Express project

3. In order to install all dependencies, we navigate into the project folder and then we run the npm install command:

```
[root@ip-172-31-95-213 node-api-postgres]# cd node-api
[root@ip-172-31-95-213 node-api]# npm install
```

4. Install pg-promise and bluebird to connect Node.js to the PostgreSQL RDS asynchronously.

The pg-promise package is a query executor for PostgreSQL databases that can be installed through npm as shown here:

```
[root@ip-172-31-95-213 node-api]# npm install pg-promise -g
[root@ip-172-31-95-213 node-api]# npm install bluebird -g
```

The bluebird package is a fully-featured promise library for Node.js that allows developers to promisify Node.js modules so that they can be used asynchronously. Promisify is a solution applied to callback functions.

We finished with setting up a Node.js project in this section. Now let's move ahead towards the next section and see how to set up a database connection from Node.js to PostgreSQL.

Setting up PostgreSQL for Node.js

We are going to set up the connection from our Node.js project to the PostgreSQL database using pg-promise, the package that we have installed recently. Under the app root directory, create the following Node.js file named queries.js:

1. The first part of the file contains require, a command to include the bluebird library, then attaches our PostgreSQL RDS connection by Node.js promise:

```
[root@ip-172-31-95-213 node-api]# vi queries.js
-------------------------------------------
var promise = require('bluebird');
var options = {
    // Initialization Options
    promiseLib: promise
};
var pgp = require('pg-promise')(options);
var connectionString =
'postgres://dba:bookdemo@atm.ck5074bwbilj.us-
east-1.rds.amazonaws.com';
var db = pgp(connectionString);
```

The preceding code uses pg-promise – a Node.js library. This library provides automatic PostgreSQL connections, PostgreSQL transactions, and the PostgreSQL query engine. The pg-promise library processes its callback chains based on promises.

We prefer to declare our promises with bluebird because bluebird is the fastest promise library for Node.js. Hence we have declared the promiseLib options of bluebird, then we next apply the promiseLib options with pg-promise to create a db connection to our AWS RDS:
'postgres://dba:bookdemo@atm.ck5074bwbilj.us-east-1.rds.amazonaws.com'.

2. The second part of the queries.js file will define the module.exports object, including query functions as Node.js modules:

```
// add query functions
module.exports =
    getAllATMLocations: getAllATMLocations,
    getSingleATMLocation: getSingleATMLocation,
    createATMLocation: createATMLocation,
    updateATMLocation: updateATMLocation,
    removeATMLocation: removeATMLocation
};
```

With the preceding module.exports statement, later on, if inside any other new Node.js script, we will send a require statement to this queries.js file, such as var db = require('/path/to/queries');, then from the new script file, we can call to five functions of the queries.js file: db.getAllATMLocations, db.getSingleATMLocation, db.createATMLocation, db.updateATMLocation, and db.removeATMLocation. Let's say we declare the following:

```
getAllATMLocations: getAllATMLocations,
```

The function name on the left-hand side of the colon (:) will be exported for the external script, such as the db object, while the module name on the right-hand side of the colon will be implemented as an internal function inside the queries.js file. These two names are not necessarily the same; you can change them to two different names.

3. The third part of this file will implement the first module to retrieve all of the ATM locations from the RDS:

```
function getAllATMLocations(req, res, next) {
        db.any('select * from public."ATM locations"')
            .then(function (data) {
                res.status(200)
                    .json({
                            status: 'success',
                            data: data,
                            message: 'Retrieved ALL
ATM locations',
                        });
                })
            .catch(function (err) {
                return next(err);
            });
    }
```

This `getAllATMLocations` module first will send a `SELECT` statement to PostgreSQL to retrieve all of the ATM locations. If the `SELECT` statement is successful, then the `success-status` value, the acquired ATM data, and a `success-message` value will be returned. Otherwise, the error exception of PostgreSQL will be caught to return.

4. As shown in the following code block, we will now implement the next module to retrieve only one specific ATM:

```
function getSingleATMLocation(req, res, next) {
        var atmID = parseInt(req.params.id);
        console.log(atmID);
        db.one('select * from public."ATM locations" where "ID" =
$1', atmID)
                .then(function (data) {
                    res.status(200)
                        .json({
                                status: 'success',
                                data: data,
                                message: 'Retrieved ONE
ATM location'
                            });
                    })
                .catch(function (err) {
                    return next(err);
                });
    }
```

This `getSingleATMLocation` module at first receives an `integer` `req.params.id` value for `atmID`, then it sends a SELECT statement to PostgreSQL to retrieve the ATM location inside the database that has the same ID value. If the SELECT statement is successful, then the success status, the acquired ATM data, and a success message will be returned. Otherwise, the error exception of PostgreSQL will be caught to return.

5. Next, we will write the following code to insert the location of the ATM.

   This is the purpose of our `insert` module: it will get inputs of an ATM's details such as bank name, address, county, city, state, and zip code with POST requests, then copy these values into an INSERT statement to add a new record into the PostgreSQL database:

```
function createATMLocation(req, res, next) {
        db.none('insert into public."ATM locations"("BankName",
"Address", "County", "City", "State", "ZipCode")' +
 'values(${BankName}, ${Address}, ${County}, ${City}, ${State},
${ZipCode})', req.body)
                .then(function () {
                        res.status(200)
                                .json({
                                        status: 'success',
                                        message: 'Inserted ONE ATM
Location'
                                });
                })
                .catch(function (err) {
                        return next(err);
                });
}
```

   If the INSERT statement is successful, then a JSON object will return with a success status and a success message, otherwise, it will catch the error exception to return.

6. Next, we will write the following code. This module will update an ATM location using the ID of the ATM:

```
function updateATMLocation(req, res, next) {
        var atmID = parseInt(req.params.id);
        console.log(atmID);
        var field = req.body.field;
        var newvalue = req.body.newvalue;
        console.log(field);
        console.log(newvalue);
```

```
                db.none('update public."ATM locations" set "'+field+'" =
\''+newvalue+'\' where "ID" = '+ atmID)
                        .then(function () {
                                res.status(200)
                                        .json({
                                                status: 'success',
                                                message: 'Updated ATM
Location'
                                        });
                        })
                        .catch(function (err) {
                                return next(err);
                        });
        }
```

7. Finally, we write the following code to delete the location of an ATM:

```
function removeATMLocation(req, res, next) {
        var atmID = parseInt(req.params.id);
        db.result('delete from public."ATM locations" where "ID" =
$1', atmID)
                .then(function (result) {
                        res.status(200)
                                .json({
                                        status: 'success',
                                        message: 'Removed an ATM
Location'
                                });
                })
                .catch(function (err) {
                        return next(err);
                });
        }
```

The `queries.js` file is now complete and the modules have been implemented. We'll now move on to working with server-side implementations.

Working with the server side: routes

In the previous section, we completed the implementation for select, insert, update, and delete functions from the database side. Now we are going to route each of the HTTP paths into these database functions, we can complete the Node.js RESTful API at this stage and then try to execute the API:

1. Let's create a route file as shown in the following code block:

```
[root@ip-172-31-95-213 node-api]# vi routes/index.js
------------------------------------------
var express = require('express');
var router = express.Router();
var path = require('path');
var db = require('../queries');

router.get('/api/atm-locations', db.getAllATMLocations);
router.get('/api/atm-locations/:id', db.getSingleATMLocation);
router.post('/api/atm-locations', db.createATMLocation);
router.put('/api/atm-locations/:id', db.updateATMLocation);
router.delete('/api/atm-locations/:id', db.removeATMLocation);
module.exports = router;
------------------------------------------
```

In the preceding code, we noticed the following elements:

- The HTTP GET request by the '/api/atm-locations' path will link to the getAllATMLocations module that is implemented inside the queries.js file.
- The POST request by the '/api/atm-locations' path will link to the createATMLocation module that is implemented inside the queries.js file.
- The PUT request by the '/api/atm-locations/:id' path will link to the updateATMLocation module that is implemented inside the queries.js file, and so on.

2. The preceding Node.js index.js file should be added to the app using the app.js file.
    1. The first part of the app.js file, as shown here, requires Node.js libraries for the app:

```
[root@ip-172-31-95-213 node-api]# npm install serve-favicon
[root@ip-172-31-95-213 node-api]# vi app.js
------------------------------------------
var createError = require('http-errors');
var express = require('express');
```

```
var path = require('path');
var cookieParser = require('cookie-parser');
var logger = require('morgan');
var indexRouter = require('./routes/index');
var usersRouter = require('./routes/users');
var favicon = require('serve-favicon');
var bodyParser = require('body-parser');
```

3. The second part, as shown in the following code block, defines a new Express app:

```
var app = express();
// view engine setup
app.set('views', path.join(__dirname, 'views'));
app.set('view engine', 'jade');
app.use(logger('dev'));
app.use(express.json());
app.use(express.urlencoded({ extended: false }));
app.use(cookieParser());
app.use(express.static(path.join(__dirname, 'public')));
app.use('/', indexRouter);
app.use('/users', usersRouter);
```

4. Finally, the following part defines error handlers for the Express app:

```
// catch 404 and forward to error handler
app.use(function(req, res, next) {
        next(createError(404));
});
// error handler
app.use(function(err, req, res, next) {
        // set locals, only providing error in development
        res.locals.message = err.message;
        res.locals.error = req.app.get('env') === 'development' ?
err : {};
        // render the error page
        res.status(err.status || 500);
        res.render('error');
});
module.exports = app;
-----------------------------------------
```

In the end, we have exported this app as the module of this `app.js` file and that completes the file.

5. Now we will start the server through the command line with the help of the following command:

```
[root@ip-172-31-95-213 node-api]# npm start
```

If you come across any errors, then execute the following command:

```
[root@ip-172-31-95-213 node-api]# npm install --save bluebird
```

The `--save` option will update the `package.json` file to list the project's dependencies. After the preceding command is executed, the console will look as shown in the following screenshot:

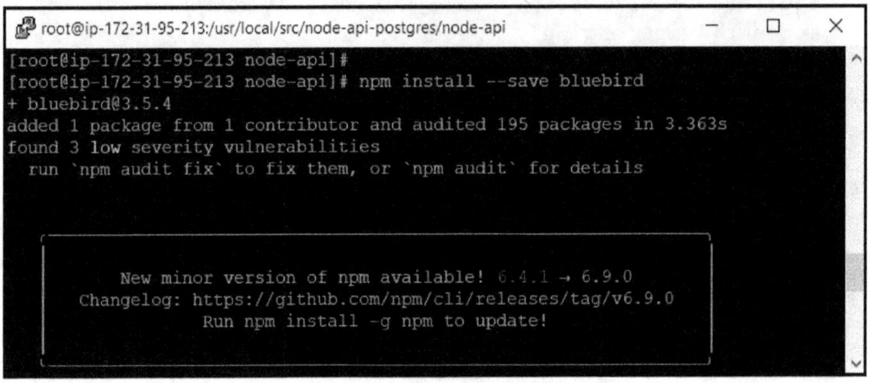

Figure 3.4 – Bluebird installation

6. As shown here, we will repeat the same command to update `pg-promise`:

```
[root@ip-172-31-95-213 node-api]# npm install --save pg-promise
```

7. Now, we will restart the server, by executing the following command:

```
[root@ip-172-31-95-213 node-api]# npm start
> node-api@0.0.0 start /usr/local/src/node-api-postgres/node-api
> node ./bin/www
```

After we execute the command to restart the server, the Node.js server starts on port 3000 as shown in the following screenshot:

```
 root@ip-172-31-95-213:/usr/local/src/node-api-postgres/node-api

[root@ip-172-31-95-213 node-api]#
[root@ip-172-31-95-213 node-api]#
[root@ip-172-31-95-213 node-api]#
[root@ip-172-31-95-213 node-api]#
[root@ip-172-31-95-213 node-api]#
[root@ip-172-31-95-213 node-api]#
[root@ip-172-31-95-213 node-api]#
[root@ip-172-31-95-213 node-api]#
[root@ip-172-31-95-213 node-api]#
[root@ip-172-31-95-213 node-api]#
[root@ip-172-31-95-213 node-api]#
[root@ip-172-31-95-213 node-api]# npm start

> node-api@0.0.0 start /usr/local/src/node-api-postgres/node-api
> node ./bin/www
```

Figure 3.5 – How to start a Node.js server

8. We have to grant access from outside into port 3000 of the Node.js server so that the API routes that we have implemented will work. You must add the correct TCP inbound rules for Node.js on the EC2 instance's security group as in *Figure 3.6*:

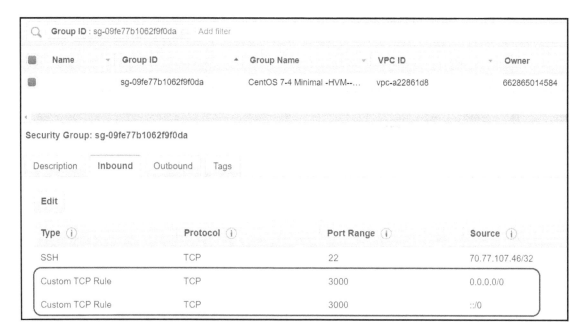

Name	Group ID	Group Name	VPC ID	Owner
	sg-09fe77b1062f9f0da	CentOS 7-4 Minimal -HVM--...	vpc-a22861d8	662865014584

Security Group: sg-09fe77b1062f9f0da

Description   **Inbound**   Outbound   Tags

Edit

Type ⓘ	Protocol ⓘ	Port Range ⓘ	Source ⓘ
SSH	TCP	22	70.77.107.46/32
Custom TCP Rule	TCP	3000	0.0.0.0/0
Custom TCP Rule	TCP	3000	::/0

Figure 3.6 – Node.js security group

9. Next, we will add the correct TCP inbound rules with the EC2 instance's private and public IPs for the PostgreSQL RDS' security group as in *Figure 3.7*:

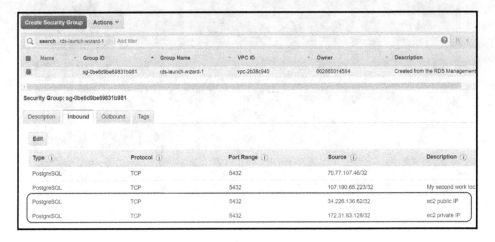

Figure 3.7 – PostgreSQL security group

10. Please include the following public DNS of the ec2. In our setup, it is from *Figure 3.8*:

```
ec2-3-209-184-46.compute-1.amazonaws.com
```

In order to copy the public DNS of your EC2 instance, please open the Amazon EC2 console at `https://console.aws.amazon.com/ec2/`:

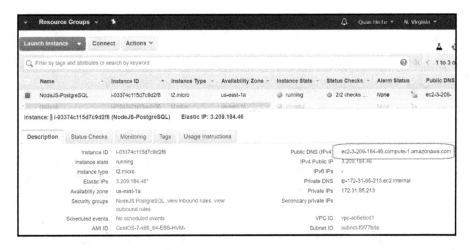

Figure 3.8 – Using the public DNS for the RESTful API

11. We will use the public DNS of the EC2 instance with port 3000 of Node.js to test our API. Please open the following web link in your browser to see if our API is working:

```
http://ec2-3-209-184-46.compute-1.amazonaws.com:3000/api/atm-
locations
```

After executing the preceding command, you will then see the JSON data for all of the ATM machines, as shown in *Figure 3.9*:

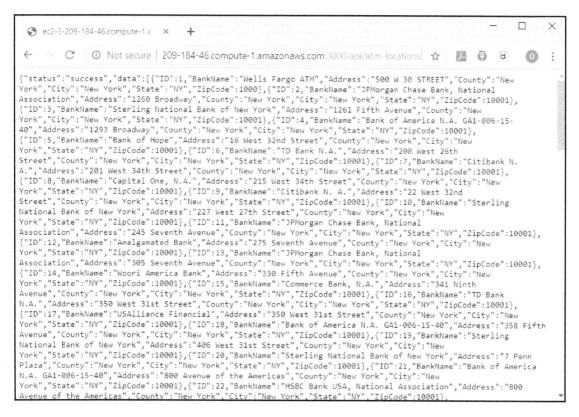

Figure 3.9 – The HTTP Get request by the RESTful API

12. Next, we test how our Node.js API will show the data of a specific ATM location of ID 111. Please try this link in your browser:

```
http://ec2-3-209-184-46.compute-1.amazonaws.com:3000/api/atm-
locations/111
```

When we run the preceding command, it will return the following output. The browser page will look as shown in the following screenshot. You will see the ATM machine with ID 111 as in *Figure 3.10*:

Figure 3.10 – A HTTP Get request for a specific ATM

13. You will have to create a duplicate putty session so that the Node.js web server keeps running, and then on the second putty, please type the following command to insert a test ATM:

```
[root@ip-172-31-95-213 node-api]# curl -X POST --data "BankName=My
Test Bank&Address=2019 Test Avenue&County=New York&City=New
York&State=NYZipCode=20190514"
http://ec2-3-209-184-46.compute-1.amazonaws.com:3000/api/atm-locati
ons
```

When we run the preceding command on the console, it will look as shown in the following screenshot. You will see the status of the successful insertion of a new ATM location as in *Figure 3.11*:

Figure 3.11 – A HTTP Post request by the RESTful API

After that, using pgAdmin, we can see the new ATM location inserted with ID 655 as shown in *Figure 3.12*:

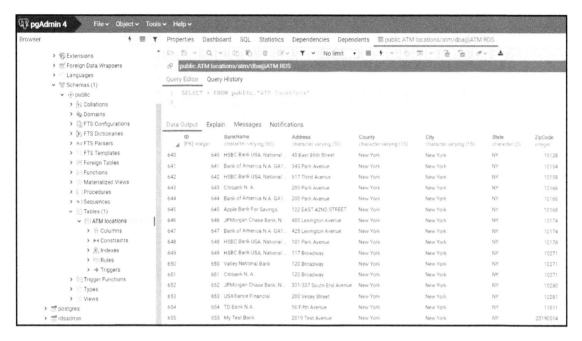

Figure 3.12 – The new inserted ATM location

14. Now, with the help of the following command, we will update the bank name of our test as shown in *Figure 3.13*:

```
[root@ip-172-31-95-213 node-api]# curl -X PUT --data
"field=BankName&newvalue=Test New ATM"
http://ec2-3-209-184-46.compute-1.amazonaws.com:3000/api/atm-locati
ons/655
```

When we run the preceding command on the console, it will look as shown in the following screenshot. You will be able to see the status of the successful update of ATM location 655:

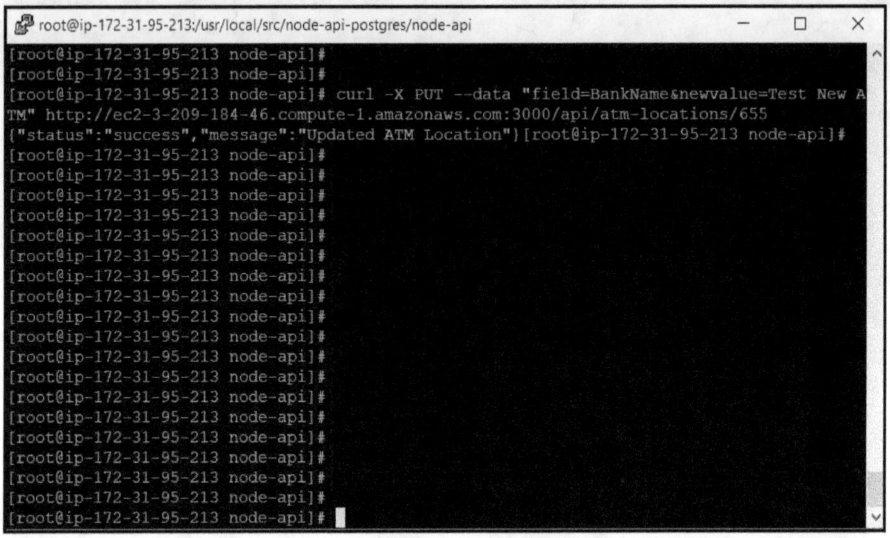

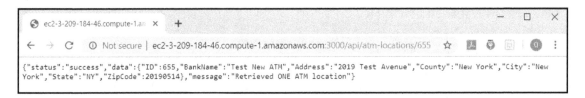

Figure 3.13 – An HTTP Put request by the RESTful API

We can GET this change on the browser as in *Figure 3.14* by using our Node.js API:

```
ec2-3-209-184-46.compute-1.ar  ×  +

←  →  C  ⓘ Not secure | ec2-3-209-184-46.compute-1.amazonaws.com:3000/api/atm-locations/655  ☆

{"status":"success","data":{"ID":655,"BankName":"Test New ATM","Address":"2019 Test Avenue","County":"New York","City":"New York","State":"NY","ZipCode":20190514},"message":"Retrieved ONE ATM location"}
```

Figure 3.14 – The updated ATM location

15. Now, with the help of the following command, we will test the deletion of the test ATM as in *Figure 3.15*:

```
[root@ip-172-31-95-213 node-api]# curl -X DELETE
http://ec2-3-209-184-46.compute-1.amazonaws.com:3000/api/atm-locati
ons/655
```

When we run the preceding command on the console, it will look as shown in the following screenshot. You will see the status of the successful DELETE request:

```
root@ip-172-31-95-213:/usr/local/src/node-api-postgres/node-api          —    □    ×
[root@ip-172-31-95-213 node-api]#
[root@ip-172-31-95-213 node-api]# curl -X DELETE http://ec2-3-209-184-46.compute
-1.amazonaws.com:3000/api/atm-locations/655
{"status":"success","message":"Removed an ATM Location"}[root@ip-172-31-95-213 n
ode-api]#
[root@ip-172-31-95-213 node-api]#
[root@ip-172-31-95-213 node-api]#
[root@ip-172-31-95-213 node-api]#
[root@ip-172-31-95-213 node-api]#
[root@ip-172-31-95-213 node-api]#
[root@ip-172-31-95-213 node-api]#
[root@ip-172-31-95-213 node-api]#
[root@ip-172-31-95-213 node-api]#
[root@ip-172-31-95-213 node-api]#
[root@ip-172-31-95-213 node-api]#
[root@ip-172-31-95-213 node-api]#
[root@ip-172-31-95-213 node-api]#
[root@ip-172-31-95-213 node-api]#
[root@ip-172-31-95-213 node-api]#
[root@ip-172-31-95-213 node-api]#
[root@ip-172-31-95-213 node-api]#
[root@ip-172-31-95-213 node-api]#
```

Figure 3.15 – An HTTP DELETE request by the RESTful API

Note: To exit the Node.js server, you can press *Ctrl + C*.

As our Node.js API has been completed, any web apps from the frontend can send their HTTP requests to make changes to the PostgreSQL database. In the next section, we are going to use Angular to create a simple frontend for our Node.js API.

In this section, we saw how to route each of the HTTP paths into these database functions. In the next section, we will create a client-side app for the Node.js API.

Working with the client side: Angular

We are going to make the client-side app for our Node.js API in this section and we will illustrate one of the advantages of Node.js by choosing a JavaScript frontend tool so that both the Node.js backend and the ongoing frontend of our system are unified because they are both written in the same JavaScript language. One of the famous frontend JavaScript tools is Angular. It is an open source framework used for building single-page web apps. Before the release of Angular, there were other tools to create dynamic web pages, but they are not as convenient as Angular:

1. First, we install Angular by using the npm command as follows:

```
[root@ip-172-31-95-213 node-api]# npm install -g @angular/cli
/usr/bin/ng -> /usr/lib/node_modules/@angular/cli/bin/ng
npm WARN optional SKIPPING OPTIONAL DEPENDENCY: fsevents@1.2.9
(node_modules/@angular/cli/node_modules/fsevents):
npm WARN notsup SKIPPING OPTIONAL DEPENDENCY: Unsupported platform
for fsevents@1.2.9: wanted {"os":"darwin","arch":"any"} (current:
{"os":"linux","arch":"x64"})
+ @angular/cli@7.3.9
added 295 packages from 180 contributors in 12.438s
```

2. Next, we will add an instance of our Angular module as shown in the following code block:

```
[root@ip-172-31-95-213 node-api]# vi public/javascripts/app.js
------------------------------------------
angular.module('nodeATM', [])
       .controller('mainController', function($scope, $http) {
              $scope.formData = {};
              $scope.atmData = {};
              // Get all atms
              $http.get('/api/atm-locations')
                      .success(function(data) {
                             $scope.atmData = data["data"];
                             console.log(data["data"]);
                      })
                      .error(function(error) {
                             console.log('Error: ' + error);
                      });
       });
------------------------------------------
```

3. We will now declare the `mainController` as shown in the following code block:

```
[root@ip-172-31-95-213 node-api]# vi public/javascripts/app.js
-----------------------------------------
angular.module('nodeATM', [])
        .controller('mainController', function($scope, $http) {
                $scope.formData = {};
                $scope.atmData = {};
```

4. As shown in the following code, we will implement the http GET method within the `mainController`:

```
        // Get all atms
        $http.get('/api/atm-locations')
                .success(function(data) {
                        $scope.atmData = data["data"];
                        console.log(data["data"]);
                })
                .error(function(error) {
                        console.log('Error: ' + error);
                });
});
```

5. Now let's update the main route in `index.js` within the "routes" folder:

```
router.get('/', (req, res, next) => {
        res.sendFile('index.html');
});
```

6. So, when the user hits the main endpoint, we send the `index.html` file. This file will contain our HTML and Angular templates. Make sure to add the following dependency as well:

```
var path = require('path');
[root@ip-172-31-95-213 node-api]# vi routes/index.js
-----------------------------------------
var express = require('express');
var router = express.Router();
var path = require('path');
var db = require('../queries');

router.get('/api/atm-locations', db.getAllATMLocations);
router.get('/api/atm-locations/:id', db.getSingleATMLocation);
router.post('/api/atm-locations', db.createATMLocation);
router.put('/api/atm-locations/:id', db.updateATMLocation);
router.delete('/api/atm-locations/:id', db.removeATMLocation);
router.get('/', function(req, res, next) {
```

```
                              res.sendFile('index.html');
        });
        module.exports = router;
        ---------------------------------------------
```

This is the screenshot for the preceding file, `routes/index.js`:

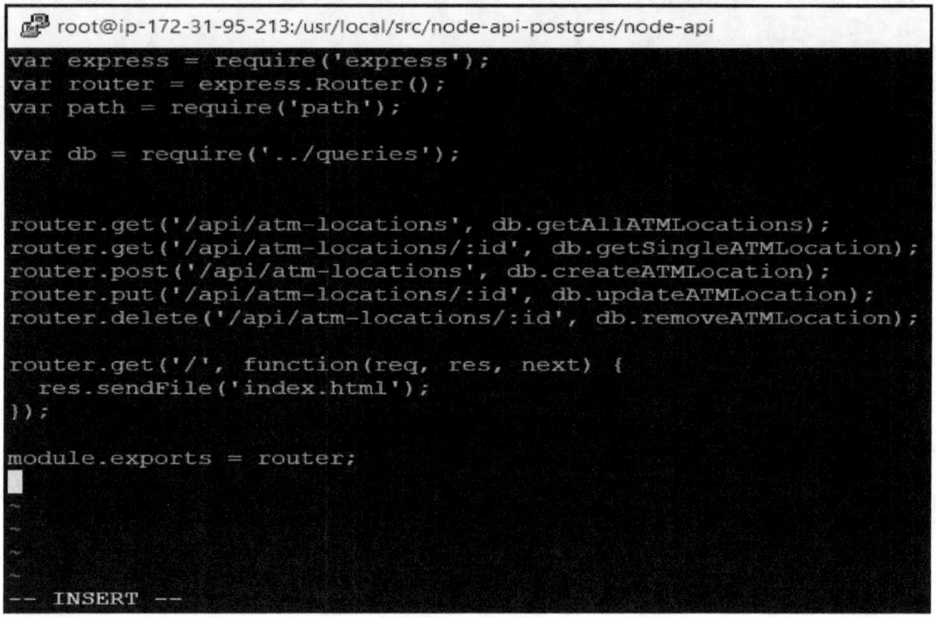

```
 root@ip-172-31-95-213:/usr/local/src/node-api-postgres/node-api

var express = require('express');
var router = express.Router();
var path = require('path');

var db = require('../queries');


router.get('/api/atm-locations', db.getAllATMLocations);
router.get('/api/atm-locations/:id', db.getSingleATMLocation);
router.post('/api/atm-locations', db.createATMLocation);
router.put('/api/atm-locations/:id', db.updateATMLocation);
router.delete('/api/atm-locations/:id', db.removeATMLocation);

router.get('/', function(req, res, next) {
  res.sendFile('index.html');
});

module.exports = router;

~
~
~
-- INSERT --
```

Figure 3.16 – Linking Angular to the Node.js RESTful API

The controller of our Angular app already links to the data model made by the Node.js API. We will move on to the next section to implement the view for our app.

Creating an Angular view

Angular is a framework for web applications by HTML and TypeScript.

The architecture of an Angular application relies on basic building blocks called NgModules to compile components. Developers use a set of NgModules to define an Angular app. Angular components define views while a view is a set of screen elements. A view is different from a service, which provides specific functionality such as dependencies to make your code modular and reusable.

A view is a defined template combining HTML, Angular directives, and binding markup together to display the web interface. Angular components of one app can define more than one view, and its multi-views can be hierarchically set up. In Angular applications, the view lives inside the HTML templates and the CSS style sheets:

1. Now, let's add our basic Angular view within `index.html` to the "public" folder. At first, the Angular application is defined by the `ng-app` directive:

```
[root@ip-172-31-95-213 node-api]# vi public/index.html
-----------------------------------------
<!DOCTYPE html>
<html ng-app="nodeATM">
        <head>
                <title>ATM App - with Node + Express + Angular +
PostgreSQL</title>
                <meta name="viewport" content="width=device-width,
initial-scale=1.0">
                <link
href="http://maxcdn.bootstrapcdn.com/bootstrap/3.3.7/css/bootstrap.
min.css" rel="stylesheet" media="screen">
        </head>
```

2. Next, as shown in the following code, we will define the Angular controller in which all the ATM bank names are shown:

```
<body ng-controller="mainController">
        <div class="container">
                <div class="header">
                        <h1>ATM App</h1>
                        <hr>
                        <h1 class="lead">Node + Express + Angular
+ PostgreSQL</h1>
                </div>
                <ul ng-repeat="atm in atmData">
                        <li>{{ atm.BankName }}</li>
                </ul>
        </div>
```

3. Finally, as shown in the following code, we will load the `angular.min.js` library before the closing tag of body:

```
                <script
src="http://code.jquery.com/jquery-2.2.4.min.js"
type="text/javascript"></script>
                <script
src="http://maxcdn.bootstrapcdn.com/bootstrap/3.3.7/js/bootstrap.mi
n.js" type="text/javascript"></script>
```

```
                    <script
src="https://cdnjs.cloudflare.com/ajax/libs/angular.js/1.5.6/angula
r.min.js"></script>
                    <script src="javascripts/app.js"></script>
        </body>
</html>
-----------------------------------------
```

After loading all the script files, we will close the body and all the other html tags of our Angular view.

4. Now, with the help of the following command, we can start our ATM app and open it in the browser as shown in *Figure 3.17*:

```
[root@ip-172-31-95-213 node-api]# npm start
> node-api@0.0.0 start /usr/local/src/node-api-postgres/node-api
> node ./bin/www
```

Because we have already routed the '/' web path to the Angular View template file /public/index.html in the previous section by the /routes/index.js file as follows:

```
router.get('/', (req, res, next) => {
        res.sendFile('index.html');
});
```

Now we open the following page inside our browser:

```
http://ec2-3-209-184-46.compute-1.amazonaws.com:3000/
```

The index.html file will execute the public/javascripts/app.js file to populate the $scope.atmData variable with all of our ATM locations from PostgreSQL and then display them in the browser as shown here:

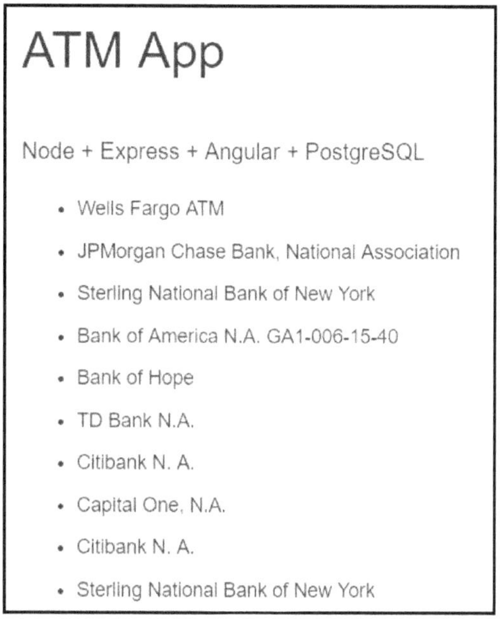

Figure 3.17 – Starting the Angular ATM app

The basic view from the preceding figure only displays all of the ATM locations, but it has not been able to add a new ATM, to update, or to delete an existing ATM location yet. Hence we are going to refactor our Angular app for these data abilities.

Angular app refactoring

This section involves the restructuring of the code written for our app. We'll do so using the following steps:

1. Let's begin by updating the module that contains the create and delete functions. We will also add new methods right after the GET method that we implemented previously in the `public/javascripts/app.js` file. Let's have a look at the following code:

```
[root@ip-172-31-95-213 node-api]# vi public/javascripts/app.js
-------------------------------------------
angular.module('nodeATM', [])
        .controller('mainController', function($scope, $http) {
                $scope.formData = {};
                $scope.atmData = {};
                // Get all atms
```

```
$http.get('/api/atm-locations')
        .success(function(data) {
                $scope.atmData = data["data"];
                console.log(data["data"]);
        })
        .error(function(error) {
                console.log('Error: ' + error);
        });
```

2. Next, as shown in the following code, we will add the new http POST method to insert a new ATM location:

```
// Create a new atm
$scope.createATM = function() {
        $http.post('/api/atm-locations', $scope.formData)
                .success(function(res) {
                        $scope.formData = {};
                        $http.get('/api/atm-locations')
                                .success(function(data) {
                                        $scope.atmData =
data["data"];

console.log(data["data"]);

                                })
                                .error(function(error) {

console.log('Error: ' + error);

                                });
                        console.log(res);
                })
                .error(function(error) {
                        console.log('Error: ' + error);
                });
};
```

3. Finally, we will add the delete function as shown in the following code:

```
// Delete an atm
$scope.deleteATM = function(atmID) {
        $http.delete('/api/atm-locations/' + atmID)
                .success(function(res) {
                        $http.get('/api/atm-locations')
                                .success(function(data) {
                                        $scope.atmData =
data["data"];

console.log(data["data"]);

                                })
```

```
                                        .error(function(error) {

console.log('Error: ' + error);

                                        });
                                console.log(res);
                        })
                        .error(function(data) {
                                console.log('Error: ' + data);
                        });
                };
});
------------------------------------------
```

The preceding methods need to be called from the Angular view, so you will have to upgrade the website so that it can include an ATM creation form and new checkboxes for ATM deletion including the ATM addresses:

1. Let's have a look and check whether changes have to be done in the view file also. The following is the first part of the view file and will remain the same as earlier:

```
[root@ip-172-31-95-213 node-api]# vi public/index.html
------------------------------------------
<!DOCTYPE html>
<html ng-app="nodeATM">
        <head>
                <title>ATM App - with Node + Express + Angular +
PostgreSQL</title>
                <meta name="viewport" content="width=device-width,
initial-scale=1.0">
                <link
href="http://maxcdn.bootstrapcdn.com/bootstrap/3.3.7/css/bootstrap.
min.css" rel="stylesheet" media="screen">
        </head>
        <body ng-controller="mainController">
                <div class="container">
                        <div class="header">
                                <h1>ATM App</h1>
                                <hr>
                                <h1 class="lead">Node + Express +
Angular + PostgreSQL</h1>
                        </div>
```

2. The middle part is added with a new web form for data insertion and by then adding many checkboxes to invoke deletion as shown here:

```
<div class="atm-form">
    <form>
        <div class="form-group">
            <input type="text"
class="form-control input-lg" placeholder="Enter Bank Name..." ng-
model="formData.BankName" style="width: 600px;margin-bottom:
10px;">
            <input type="text"
class="form-control input-lg" placeholder="Enter Address..." ng-
model="formData.Address" style="width:600px;margin-bottom: 10px;">
            <input type="text"
class="form-control input-lg" placeholder="Enter County..." ng-
model="formData.County" style="width:300px;margin-bottom: 10px;">
            <input type="text"
class="form-control input-lg" placeholder="Enter City..." ng-
model="formData.City" style="width:300px;margin-bottom: 10px;">
            <input type="text"
class="form-control input-lg" placeholder="Enter State..." ng-
model="formData.State" style="width:100px;margin-bottom: 10px;">
            <input type="text"
class="form-control input-lg" placeholder="Enter ZipCode..." ng-
model="formData.ZipCode" style="width:200px;margin-bottom: 10px;">
        </div>
        <button type="submit"
class="btn btn-primary btn-lg btn-block" ng-click="createATM()"
style="width:300px;margin-bottom: 10px;">Add ATM</button>
    </form>
</div>
<br>
<ul ng-repeat="atm in atmData">
    <li><h3><input class="lead"
type="checkbox" ng-click="deleteATM(atm.ID)">{{ atm.BankName }}: {{
atm.Address }}, {{ atm.City }}, {{ atm.State }} {{atm.ZipCode
}}</h3></li>
    <hr>
</ul>
</div>
```

3. The final part of the view file, as shown here, remains the same as this part only contains external scripts:

```
<script
src="http://code.jquery.com/jquery-2.2.4.min.js"
type="text/javascript"></script>
<script
```

```
src="http://maxcdn.bootstrapcdn.com/bootstrap/3.3.7/js/bootstrap.mi
n.js" type="text/javascript"></script>
                <script
src="https://cdnjs.cloudflare.com/ajax/libs/angular.js/1.5.6/angula
r.min.js"></script>
                <script src="javascripts/app.js"></script>
        </body>
    </html>
----------------------------------------
```

We have made improvements to our view file now.

1. Now our Angular app has been refactored, we'll next start the app and see how it looks in the browser:

```
[root@ip-172-31-95-213 node-api]# npm start
> node-api@0.0.0 start /usr/local/src/node-api-postgres/node-api
> node ./bin/www
```

We can see our new form for inserting the ATM location shown in *Figure 3.18*:

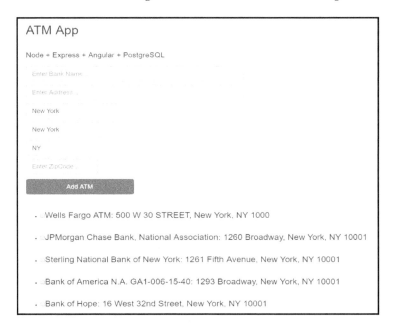

Figure 3.18 – The final Angular app

After the form, each ATM has a checkbox added to invoke deletion ability.

2. Let's try to add a test bank, then click the **Add ATM** button at the end of the form as in *Figure 3.19*:

Figure 3.19 – Adding a test ATM

3. Then, please scroll down to the end of the web page. You will see the newly added bank as in *Figure 3.20*:

Figure 3.20 – The ATM newly added by Angular

4. Please select the checkbox of **TEST BANK** to remove it. You should see **TEST BANK** will be removed immediately from the web page.

The basic Angular client app has been created with viewing, insertion, updating, and deletion abilities for our ATM locations. Going ahead in the next section, we will test our JavaScript (Node.js and Angular) automatically with Mocha and Chai.

Automated testing with Mocha and Chai

Mocha and Chai are usually used for JavaScript unit tests. Mocha and Chai run on Node. js as well as in the browser. Mocha is a test framework and Chai is a BDD / TDD assertion library. We will be using them to design our automated test scripts:

1. First, let's install Mocha and Chai with the help of the following command:

    ```
    [root@ip-172-31-95-213 node-api]# npm install --global mocha
    [root@ip-172-31-95-213 node-api]# npm install --save-dev mocha
    [root@ip-172-31-95-213 node-api]# npm install --save-dev chai
    [root@ip-172-31-95-213 node-api]# npm install --save-dev chai-http
    ```

2. Next, we will install the `request` package to send our test HTTP requests as shown here:

    ```
    [root@ip-172-31-95-213 node-api]# npm install request --save-dev
    ```

3. We have to include the Mocha `test` command in the `scripts` object inside the `package.json` file:

    ```
    [root@ip-172-31-95-213 node-api]# vi package.json
    ----------------------------------------
      {
              "name": "node-api",
              "version": "0.0.0",
              "private": true,
              "scripts": {
                      "start": "node ./bin/www",
                      "test": "mocha --timeout 10000"
              },
    ```

After we have added the unit tests' timeout as `10000`, our Mocha will set a timeout of 10 seconds. The rest of the file remains the same:

    ```
        "dependencies": {
                "bluebird": "^3.5.4",
                "cookie-parser": "~1.4.4",
    ```

```
        "debug": "~2.6.9",
        "express": "~4.16.1",
        "http-errors": "~1.6.3",
        "jade": "~1.11.0",
        "morgan": "~1.9.1",
        "pg-promise": "^8.7.2",
        "serve-favicon": "^2.5.0"
    },
    "devDependencies": {
        "chai": "^4.2.0",
        "chai-http": "^4.3.0",
        "mocha": "^6.1.4",
        "request": "^2.88.0"
    }
}
```

----

4. Please edit the tests by Mocha and Chai. At first, we require the `chai` library and the `chai-http` plugin as shown in the following code block:

```
[root@ip-172-31-95-213 node-api]# mkdir test
[root@ip-172-31-95-213 node-api]# vi test/test.js
-------------------------------------------
let chai = require('chai');
let chaiHttp = require('chai-http');
let server = require('../app');
let should = chai.should();
chai.use(chaiHttp);
```

5. We will then describe the test for `/GET` all ATM locations as shown in the following code:

```
describe('/GET atm', () => {
    it('it should GET all the ATMs', (done) => {
        chai.request(server)
            .get('/api/atm-locations')
            .end((err, res) => {
                res.should.have.status(200);
                let data = res.body["data"];
                let rstatus = res.body["status"];
                rstatus.should.be.eql('success');
                data.should.be.a('array');
                done();
            });
    });
});
```

6. As shown in the following code block, we will describe a test for /POST a new ATM location:

```
/*
 * Test the /POST route
 */
describe('/POST atm', () => {
    it('it should POST an atm', (done) => {
        let atm = {
            BankName: "TEST BANK",
            Address: "999 Test Blvd",
            County: "New York",
            City: "New York",
            State: "NY",
            ZipCode: 999999
        }
        chai.request(server)
            .post('/api/atm-locations')
            .send(atm)
            .end((err, res) => {
                res.should.have.status(200);
                let rstatus = res.body["status"];
                let rmessage = res.body["message"];

                rstatus.should.be.eql('success');
                rmessage.should.be.eql('Inserted ONE ATM
Location');
                done();
            });
    });
});
```

7. Next, we will write a test for /GET a specific ATM location as shown in the following code block:

```
/*
 * Test GET an atm
 */
describe('/GET one atm', () => {
    it('it should GET the new added atm', (done) => {
        chai.request(server)
            .get('/api/atm-locations/640')
            .end((err, res) => {
                let atm = {
                    ID: 640,
                    BankName: "HSBC Bank USA, National
Association",
                    Address: "45 East 89th Street",
```

```
                            County: "New York",
                            City: "New York",
                            State: "NY",
                            ZipCode: 10128
                        }
                        res.should.have.status(200);
                        let data = res.body["data"];
                        let rstatus = res.body["status"];
                        let rmessage = res.body["message"];
                        data.should.be.eql(atm);
                        rstatus.should.be.eql('success');
                        rmessage.should.be.eql('Retrieved ONE ATM
location');
                        done();
                    });
            });
        });
```

8. As shown in the following code block, we will write a test for /DELETE a specific ATM location using its ID:

```
/*
 * Delete an atm
 */
describe('/DELETE atm', () => {
    it('it should DELETE the new added atm', (done) => {
        chai.request(server)
            .delete('/api/atm-locations/655')
            .end((err, res) => {
                res.should.have.status(200);
                let rstatus = res.body["status"];
                let rmessage = res.body["message"];

                rstatus.should.be.eql('success');
                rmessage.should.be.eql('Removed an ATM Location');
                done();
            });
    });
});
------------------------------------------
```

9. Now we will start running our tests with the help of the following command:

```
[root@ip-172-31-95-213 node-api]# npm test
> node-api@0.0.0 test /usr/local/src/node-api-postgres/node-api
> mocha --timeout 10000
    /GET atm
 GET /api/atm-locations 200 80.219 ms - 96695
```

```
            ✓ it should GET all the ATMs (110ms)
       /POST atm
   POST /api/atm-locations 200 13.227 ms - 58
            ✓ it should POST an atm
       /GET one atm
   640
   GET /api/atm-locations/640 200 3.288 ms - 225
            ✓ it should GET the new added atm
       /DELETE atm
   DELETE /api/atm-locations/655 200 1.478 ms - 72
            ✓ it should DELETE the new added atm

   4 passing (146ms)
```

The text execution on the console will look as shown in *Figure 3.21*:

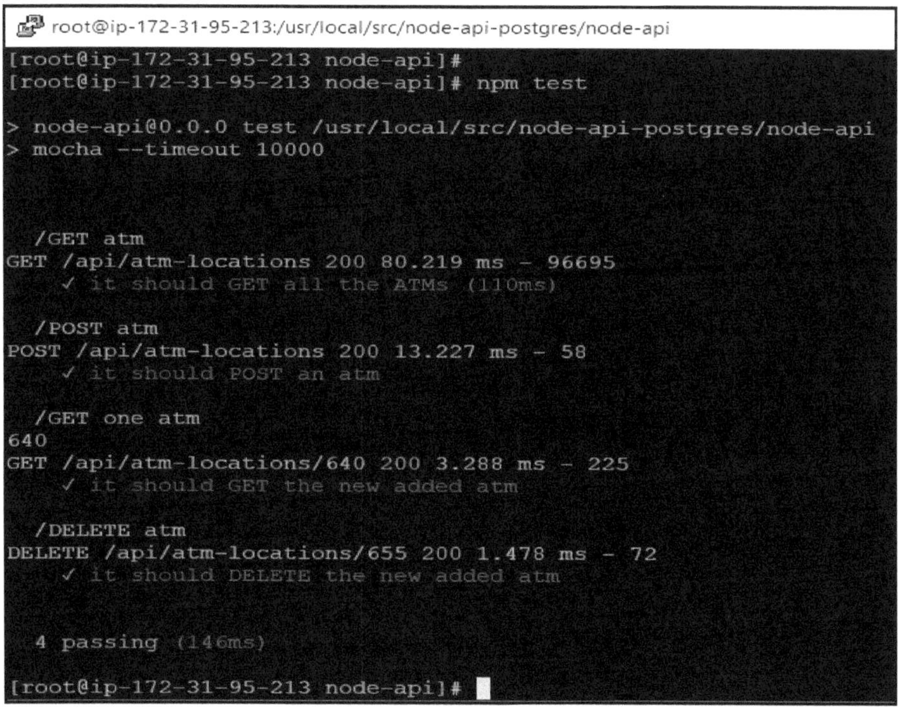

Figure 3.21 – Mocha and Chai tests

All of the /GET (all ATMs), /POST,, /GET (one ATM), and /DELETE tests have been
successful, so Mocha and Chai are very good for automated test scripts for any Node.js
APIs.

Summary

In this chapter, we started by using an AWS EC2 instance to create an API from Node.js to PostgreSQL. We practiced installing Node.js and created an Express Node.js app. Then we learned how to set up a PostgreSQL database connection with Node.js. Going ahead, we set up API routes in Node.js, the Angular module, and the controller. We learned how to set up an Angular view and re-structured the app. Finally, we learned how to create a BDD/TDD library with Mocha and Chai.

In the next chapter, we will demonstrate how to install and configure PostgreSQL to use it with a Django application.

4

Managing Bank ATM Locations Using PostgreSQL and Django

In this chapter, you will learn how to install and configure PostgreSQL so that you can use it with a Django application. First, we will install the required software. Then, we will create database credentials for our application. Finally, we will start and configure a new Django project that uses PostgreSQL as the backend.

The project of this chapter will show you how to implement a banking system to manage ATM locations within a typical city through the Django environment on the PostgreSQL 12 **Relational Database Service (RDS)** from **Amazon Web Services (AWS)**, which we explored in Chapter 3, *Using PostgreSQL and NodeJS for Banking Transactions*.

In this chapter, we will cover the following topics:

- Setting up a Django project
- Django database settings with PostgreSQL
- Database models in Django
- Migrating the database
- Understanding the Django user interface – admin, views, templates, and URLs

Technical requirements

This chapter will take developers between 10 and 12 hours to develop an ATM Django app.

The code files for this chapter are available at `https://github.com/PacktPublishing/` `Developing-Modern-Database-Applications-with-PostgreSQL/tree/master/Chapter04`.

Setting up a Django project

We will use the same CentOS 7 EC2 instance in our AWS account from `Chapter 3`, *Using PostgreSQL and NodeJS for Banking Transactions*, to deploy our Django project. So we put our EC2 instance with the built-in user for our current CentOS AMI user:

```
[centos@ip-172-31-95-213 ~]$ sudo su
[root@ip-172-31-95-213 centos]# cd /usr/local/src/
```

Following this `sudo` command, the rest of our statements are going to be executed under the user root.

Installing Python 3

By default, CentOS 7 comes with Python 2.7.5, and Python 2.7.5 is a crucially built-in part of the CentOS basesystem. However, we intend to create our project with Python version 3.7, so we will now install Python 3.7 on CentOS 7.

At the time of writing this book, Python 3.7 is the current version, and its installation requires the GCC compiler on our `ec2` instance. Please refer to the following steps to install the prerequisites for Python before installing it:

1. First, we enable SCL by installing the CentOS SCL release file. This is included in the CentOS Extras repository:

   ```
   [root@ip-172-31-95-213 src]# yum install gcc openssl-devel bzip2-
   devel libffi-devel
   ```

2. After executing the preceding command, the console will look similar to *Figure 4.1*. Please answer `y` and you should see something similar to the following screenshot:

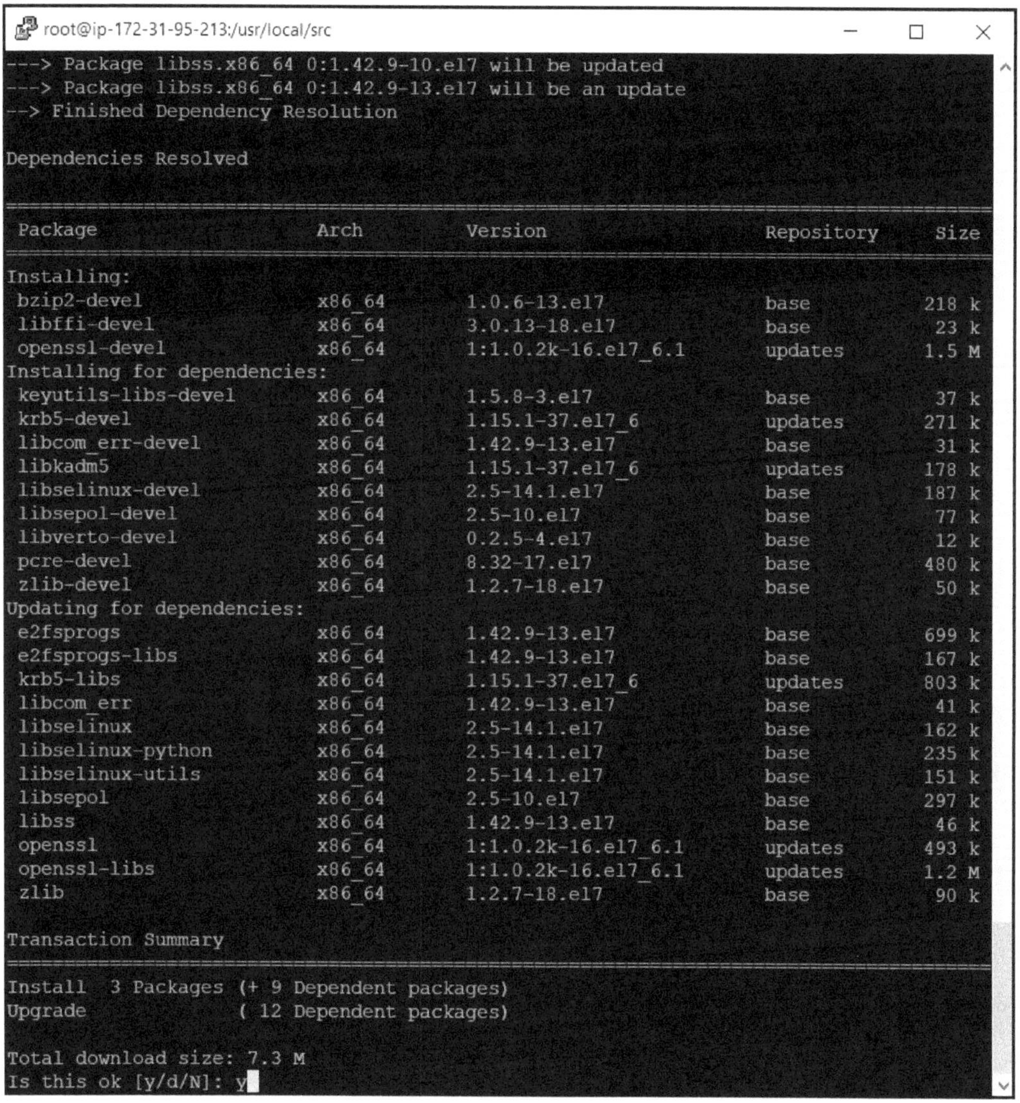

Figure 4.1 – Installing the CentOS SCL release file

3. Next, download Python 3.7 using the following command from Python's official website:

```
[root@ip-172-31-95-213 src]# wget
https://www.python.org/ftp/python/3.7.3/Python-3.7.3.tgz
[root@ip-172-31-95-213 src]# tar xzf Python-3.7.3.tgz
[root@ip-172-31-95-213 src]# cd Python-3.7.3
```

```
[root@ip-172-31-95-213 Python-3.7.3]# ./configure --enable-
optimizations
[root@ip-172-31-95-213 Python-3.7.3]# make altinstall
```

4. Now remove the downloaded source archive file from your system:

```
[root@ip-172-31-95-213 Python-3.7.3]# rm
/usr/local/src/Python-3.7.3.tgz
rm: remove regular file '/usr/local/src/Python-3.7.3.tgz'? y
[root@ip-172-31-95-213 Python-3.7.3]#
```

5. As shown in the following code block, check the version of Python that has been installed. Use the `python3.7` command instead of just `python`:

```
[root@ip-172-31-95-213 Python-3.7.3]# python3.7 -V
 Python 3.7.3
[root@ip-172-31-95-213 Python-3.7.3]# cd ..
[root@ip-172-31-95-213 src]#
```

Now that we have successfully installed Python 3.7, we can move on to create a virtual environment for our Django application.

Creating a virtual environment

One of the biggest difficulties with Django developers is that a project created in one Django version is not compatible with another project in a different Django version. A virtual environment is a Python tool that will create a new environment. Each virtual environment is project-specific and isolated with its own set of packages, settings, and dependencies. Therefore, a virtual environment is one solution to resolve the version difficulty of Django projects. The steps to create a virtual environment are as follows:

1. First, create a separate folder for our Django work, as follows:

```
[root@ip-172-31-95-213 src]# mkdir django_app
[root@ip-172-31-95-213 src]# cd django_app
[root@ip-172-31-95-213 django_app]
```

2. Next, run the following command to create a new virtual environment:

```
[root@ip-172-31-95-213 django_app]# pip python3.7 -m venv venv
```

Following the execution of the preceding command, a directory called `venv` will be created. It contains a copy of the Python binary, the `pip` package manager, the standard Python library, and other supporting files. You can set the name of the virtual environment to any name of your choice:

```
[root@ip-172-31-95-213 django_app]# ls
venv
```

3. The virtual environment will start after you run the following script:

```
[root@ip-172-31-95-213 django_app]# source venv/bin/activate
```

4. After the virtual environment has been activated, the virtual environment's `bin` directory will appear at the start of the `$PATH` variable. The shell's prompt will display the name of the virtual environment that you're currently using. In our case, that is `venv`. Let's take a look at the following command:

```
(venv) [root@ip-172-31-95-213 django_app]# echo $PATH
```

After executing the preceding command, if your console looks similar to the following screenshot, this means that you have successfully created your virtual environment:

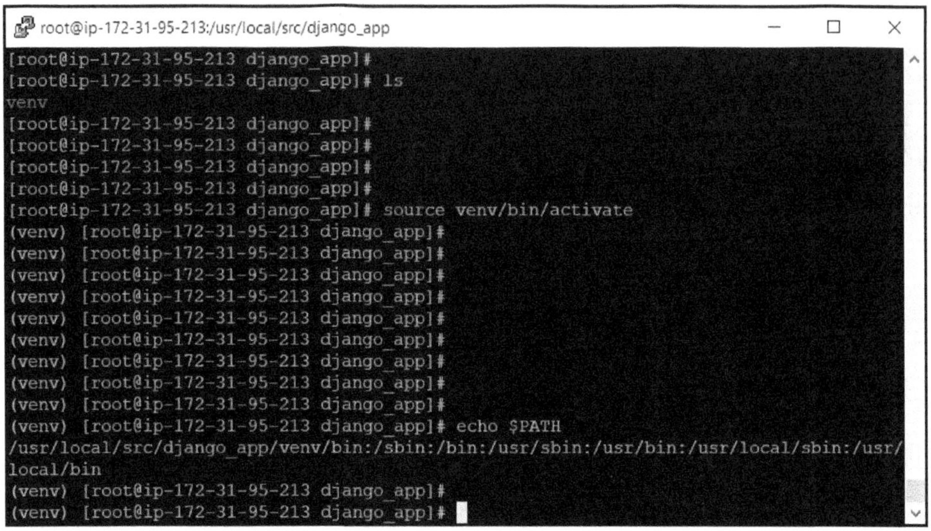

Figure 4.2 – Activating the virtual environment

To exit the virtual environment, you can just type in `deactivate`, which sets the system back to what it was earlier.

Installing Django

The virtual environment has now been activated. Therefore, we can use the Python package manager, `pip`, to install Django. Perform the following steps:

1. Install Django using the following command:

    ```
    (venv) [root@ip-172-31-95-213 django_app]# pip3 install django
    ```

   The following screenshot shows the installation output on our console:

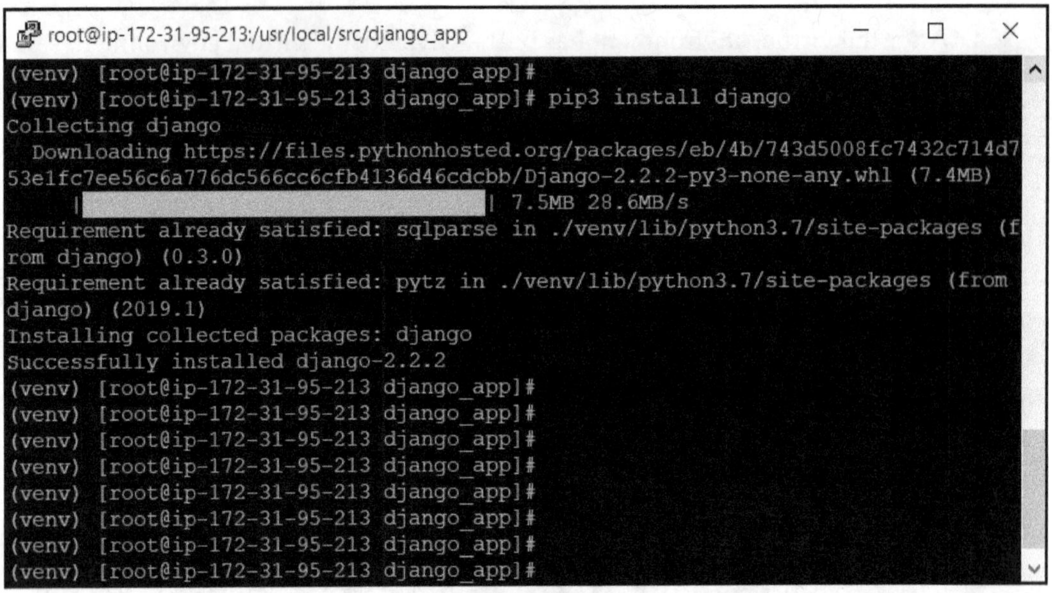

   ```
    root@ip-172-31-95-213:/usr/local/src/django_app          —    □    ×

   (venv) [root@ip-172-31-95-213 django_app]#
   (venv) [root@ip-172-31-95-213 django_app]# pip3 install django
   Collecting django
     Downloading https://files.pythonhosted.org/packages/eb/4b/743d5008fc7432c714d7
   53e1fc7ee56c6a776dc566cc6cfb4136d46cdcbb/Django-2.2.2-py3-none-any.whl (7.4MB)
         |                                          | 7.5MB 28.6MB/s
   Requirement already satisfied: sqlparse in ./venv/lib/python3.7/site-packages (f
   rom django) (0.3.0)
   Requirement already satisfied: pytz in ./venv/lib/python3.7/site-packages (from
   django) (2019.1)
   Installing collected packages: django
   Successfully installed django-2.2.2
   (venv) [root@ip-172-31-95-213 django_app]#
   (venv) [root@ip-172-31-95-213 django_app]#
   (venv) [root@ip-172-31-95-213 django_app]#
   (venv) [root@ip-172-31-95-213 django_app]#
   (venv) [root@ip-172-31-95-213 django_app]#
   (venv) [root@ip-172-31-95-213 django_app]#
   (venv) [root@ip-172-31-95-213 django_app]#
   (venv) [root@ip-172-31-95-213 django_app]#
   ```

   Figure 4.3 – Installing Django

2. The following command will help you to verify the installation process. It will print the Django version:

    ```
    (venv) [root@ip-172-31-95-213 django_app]# python3.7 -m django --
    version
       2.2.6
    ```

3. Surprisingly, the installation of Django is very simple and quick for developers. Version 2.2.6 of Django requires SQLite versions later than SQLite 3.8.3; hence, you will have to install SQLite 3.29.0 as follows:

    ```
    (venv) [root@ip-172-31-95-213 django_app]# cd ..
    (venv) [root@ip-172-31-95-213 src]# wget
    ```

```
https://www.sqlite.org/2019/sqlite-autoconf-3290000.tar.gz
(venv) [root@ip-172-31-95-213 src]# tar zxvf sqlite-
autoconf-3290000.tar.gz
(venv) [root@ip-172-31-95-213 src]# cd sqlite-autoconf-3290000
(venv) [root@ip-172-31-95-213 sqlite-autoconf-3290000]# ./configure
--prefix=$HOME/opt/sqlite
(venv) [root@ip-172-31-95-213 sqlite-autoconf-3290000]# make &&
make install
(venv) [root@ip-172-31-95-213 sqlite-autoconf-3290000]# export
PATH=$HOME/opt/sqlite/bin:$PATH
(venv) [root@ip-172-31-95-213 sqlite-autoconf-3290000]# export
LD_LIBRARY_PATH=$HOME/opt/sqlite/lib
(venv) [root@ip-172-31-95-213 sqlite-autoconf-3290000]# export
LD_RUN_PATH=$HOME/opt/sqlite/lib
(venv) [root@ip-172-31-95-213 sqlite-autoconf-3290000]# source
~/.bash_profile
(venv) [root@ip-172-31-95-213 sqlite-autoconf-3290000]# sqlite3 --
version
3.29.0 2019-07-10 17:32:03
fc82b73eaac8b36950e527f12c4b5dc1e147e6f4ad2217ae43ad82882a88bfa6
(venv) [root@ip-172-31-95-213 sqlite-autoconf-3290000]# cd ..
(venv) [root@ip-172-31-95-213 src]# rm sqlite-
autoconf-3290000.tar.gz
rm: remove regular file 'sqlite-autoconf-3290000.tar.gz'? y
(venv) [root@ip-172-31-95-213 src]# cd django_app
(venv) [root@ip-172-31-95-213 django_app]#
```

Now you can continue to use Django version 2.2.6 without any problems.

Creating a Django project

In order to create a Django project, we have to autogenerate some code that establishes a Django project that is a collection of settings, including the database configurations, Django-specific options, and application-specific settings. Perform the following steps:

1. We will start by running the following command:

    ```
    (venv) [root@ip-172-31-95-213 django_app]# django-admin
    startproject atmproject
    ```

    The preceding command will create an `atmproject` directory in your current directory with the following structure:

    ```
    (venv) [root@ip-172-31-95-213 django_app]# yum install tree -y
    (venv) [root@ip-172-31-95-213 django_app]# tree atmproject
     atmproject
    ```

```
├──── atmproject
│      ├──── __init__.py
│      ├──── settings.py
│      ├──── urls.py
│      └──── wsgi.py
└──── manage.py
```

2. You have successfully created a new Django project, as shown in the following screenshot:

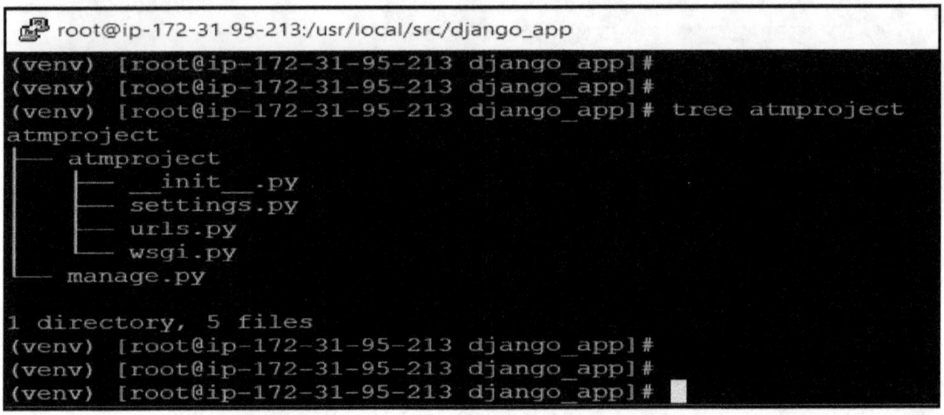

Figure 4.4 – The Django project structure

The preceding project, that is, our environment, has been set up. In the next section, we will learn how to create an app.

Creating the ATM app

Each app that we write in Django consists of a Python package, and we can use Django to automatically generate the basic app directory structure. One Django project can include many apps and configurations for a particular website, and an app can be included inside multiple projects.

Let's start by creating our own ATM app with the help of the following command:

```
(venv) [root@ip-172-31-95-213 django_app]# cd atmproject
(venv) [root@ip-172-31-95-213 atmproject]# python3.7 manage.py startapp
atmapp
```

The following is the structure of an app:

```
(venv) [root@ip-172-31-95-213 atmproject]# tree atmapp
atmapp
├── admin.py
├── apps.py
├── __init__.py
├── migrations
│   └── __init__.py
├── models.py
├── tests.py
└── views.py
```

In the console, the preceding structure will appear as follows:

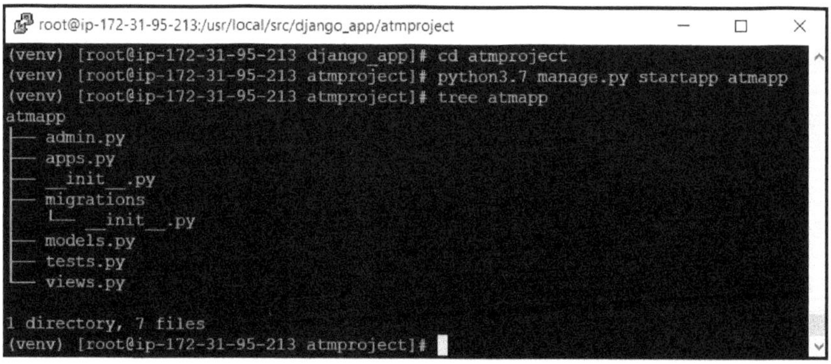

Figure 4.5 – The Django app structure

Now that we have generated our Django project and ATM app structures, next, we are going to arrange our PostgreSQL RDS settings for this project.

Django database settings with PostgreSQL

By default, Django uses the SQLite database. For production applications, you can use the PostgreSQL, MariaDB, Oracle, or MySQL databases. For this book, it will be PostgreSQL. Perform the following steps:

1. The virtual environment is active; now, we will install the `psycopg2` PostgreSQL adaptor with the help of the following command:

```
(venv) [root@ip-172-31-95-213 atmproject]# pip3 install psycopg2-
binary
```

After the preceding command has been executed, the console should appear as follows:

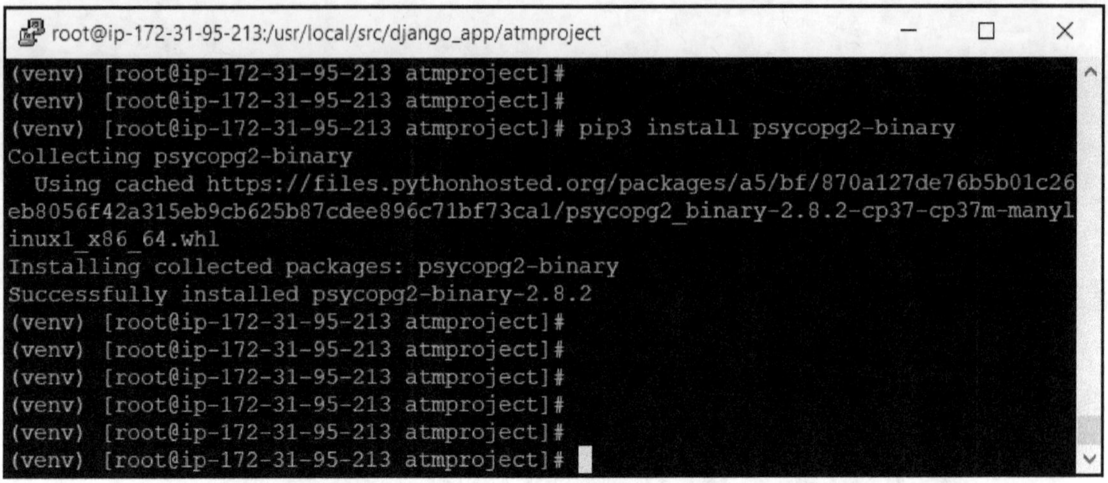

```
root@ip-172-31-95-213:/usr/local/src/django_app/atmproject                    —   □   ×
(venv) [root@ip-172-31-95-213 atmproject]#
(venv) [root@ip-172-31-95-213 atmproject]#
(venv) [root@ip-172-31-95-213 atmproject]# pip3 install psycopg2-binary
Collecting psycopg2-binary
  Using cached https://files.pythonhosted.org/packages/a5/bf/870a127de76b5b01c26
eb8056f42a315eb9cb625b87cdee896c71bf73ca1/psycopg2_binary-2.8.2-cp37-cp37m-manyl
inux1_x86_64.whl
Installing collected packages: psycopg2-binary
Successfully installed psycopg2-binary-2.8.2
(venv) [root@ip-172-31-95-213 atmproject]#
(venv) [root@ip-172-31-95-213 atmproject]#
(venv) [root@ip-172-31-95-213 atmproject]#
(venv) [root@ip-172-31-95-213 atmproject]#
(venv) [root@ip-172-31-95-213 atmproject]#
(venv) [root@ip-172-31-95-213 atmproject]#
```

Figure 4.6 – The psycopg2 PostgreSQL adapter

2. Now, open the `settings` file of the project to declare the database's RDS endpoint from AWS along with the username, password, and database name, as follows:

```
(venv) [root@ip-172-31-95-213 atmproject]# vi
atmproject/settings.py
-------------------------------------------
DATABASES = {
 'default': {
   'ENGINE': 'django.db.backends.postgresql_psycopg2',
   'NAME': 'atm',
   'USER': 'dba',
   'PASSWORD': 'bookdemo',
   'HOST': 'atm.ck5074bwbilj.us-east-1.rds.amazonaws.com',
   'PORT': '5432',
  }
}
-------------------------------------------
```

3. After the preceding command has been executed, the console should look similar to the following screenshot. Now you must save and close the file:

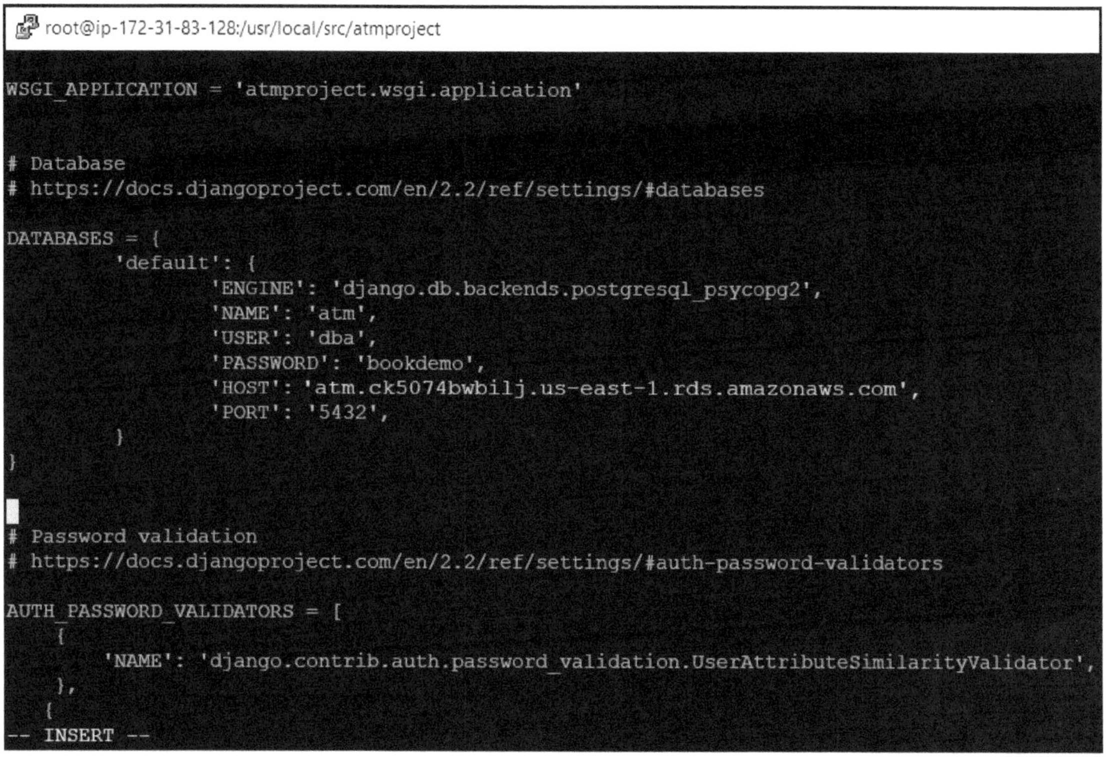

```
root@ip-172-31-83-128:/usr/local/src/atmproject

WSGI_APPLICATION = 'atmproject.wsgi.application'


# Database
# https://docs.djangoproject.com/en/2.2/ref/settings/#databases

DATABASES = {
        'default': {
                'ENGINE': 'django.db.backends.postgresql_psycopg2',
                'NAME': 'atm',
                'USER': 'dba',
                'PASSWORD': 'bookdemo',
                'HOST': 'atm.ck5074bwbilj.us-east-1.rds.amazonaws.com',
                'PORT': '5432',
        }
}


# Password validation
# https://docs.djangoproject.com/en/2.2/ref/settings/#auth-password-validators

AUTH_PASSWORD_VALIDATORS = [
    {
        'NAME': 'django.contrib.auth.password_validation.UserAttributeSimilarityValidator',
    },
    {
-- INSERT --
```

Figure 4.7 – The PostgreSQL settings inside Django

4. Next, let's create a database schema for this app with the help of the following command:

```
(venv) [root@ip-172-31-95-213 atmproject]# python3.7 manage.py
migrate
```

Once the preceding command has been executed, the console should appear as follows:

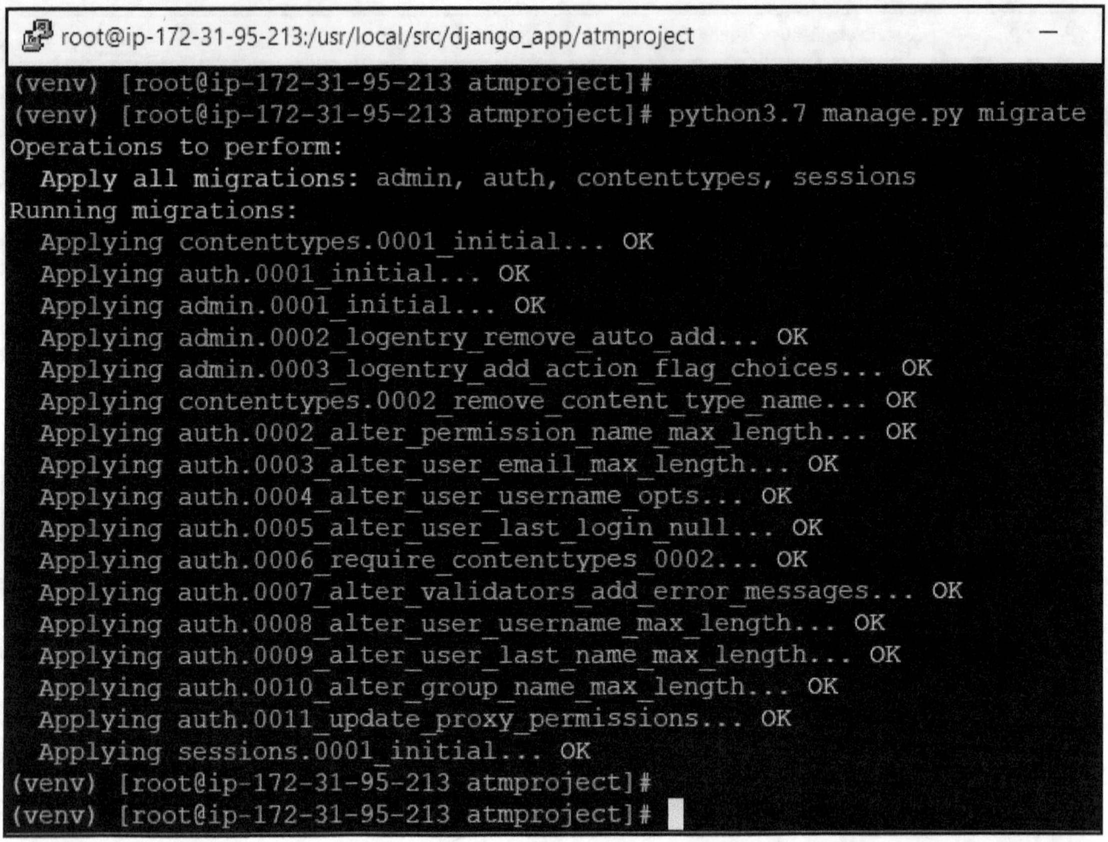

```
root@ip-172-31-95-213:/usr/local/src/django_app/atmproject                       —
(venv) [root@ip-172-31-95-213 atmproject]#
(venv) [root@ip-172-31-95-213 atmproject]# python3.7 manage.py migrate
Operations to perform:
  Apply all migrations: admin, auth, contenttypes, sessions
Running migrations:
  Applying contenttypes.0001_initial... OK
  Applying auth.0001_initial... OK
  Applying admin.0001_initial... OK
  Applying admin.0002_logentry_remove_auto_add... OK
  Applying admin.0003_logentry_add_action_flag_choices... OK
  Applying contenttypes.0002_remove_content_type_name... OK
  Applying auth.0002_alter_permission_name_max_length... OK
  Applying auth.0003_alter_user_email_max_length... OK
  Applying auth.0004_alter_user_username_opts... OK
  Applying auth.0005_alter_user_last_login_null... OK
  Applying auth.0006_require_contenttypes_0002... OK
  Applying auth.0007_alter_validators_add_error_messages... OK
  Applying auth.0008_alter_user_username_max_length... OK
  Applying auth.0009_alter_user_last_name_max_length... OK
  Applying auth.0010_alter_group_name_max_length... OK
  Applying auth.0011_update_proxy_permissions... OK
  Applying sessions.0001_initial... OK
(venv) [root@ip-172-31-95-213 atmproject]#
(venv) [root@ip-172-31-95-213 atmproject]# █
```

Figure 4.8 – Creating the Django database schema

5. Now open **pgAdmin**; you will see that 10 more tables have been added to the RDS, as shown in the following screenshot:

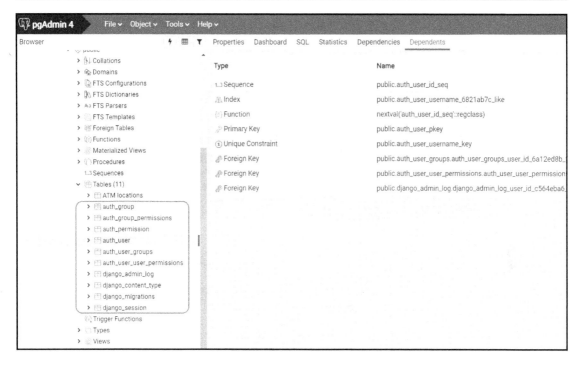

Figure 4.9 – Django tables

6. As shown in the following script, we need to create a Django superuser for the app:

```
(venv) [root@ip-172-31-95-213 atmproject]# python3.7 manage.py
createsuperuser
```

Please fill in these values as follows:

- **Username** = `admin`.
- **Email address** = (this is your email).
- The password requires eight characters; for example, here, we can type in `djangoapp`.

After the preceding command has been executed, the console should look similar to the following screenshot:

```
root@ip-172-31-95-213:/usr/local/src/django_app/atmproject                    —    □    ×
(venv) [root@ip-172-31-95-213 atmproject]#
(venv) [root@ip-172-31-95-213 atmproject]#
(venv) [root@ip-172-31-95-213 atmproject]# python3.7 manage.py createsuperuser
Username (leave blank to use 'root'): admin
Email address: postgresql.ha@gmail.com
Password:
Password (again):
Superuser created successfully.
(venv) [root@ip-172-31-95-213 atmproject]#
(venv) [root@ip-172-31-95-213 atmproject]#
(venv) [root@ip-172-31-95-213 atmproject]#
(venv) [root@ip-172-31-95-213 atmproject]#
```

Figure 4.10 – The Django superuser

This user is not the PostgreSQL master user for the RDS. While the database username is used by your app to access the database, the Django superuser is used to log into the Django admin site.

7. Please set up the inbound rule to allow access to your Django site at port 8000 from any browsers, as shown in the following screenshot:

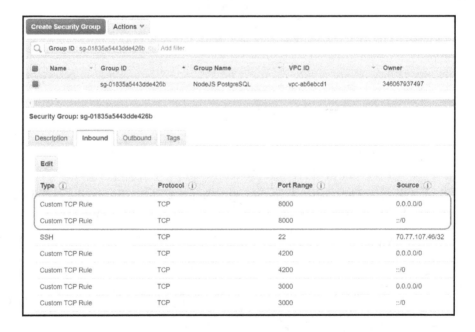

Figure 4.11 – The Django web security group

8. As our EC2 instance's IP address is `3.209.184.46` and the public DNS for this EC2 instance is `ec2-3-209-184-46.compute-1.amazonaws.com`, add the following two lines of code into the Django project settings:

```
(venv) [root@ip-172-31-95-213 atmproject]# vi
atmproject/settings.py
-------------------------------------------
ALLOWED_HOSTS = ['3.209.184.46',
'ec2-3-209-184-46.compute-1.amazonaws.com']
AUTHENTICATION_BACKENDS = (
        ('django.contrib.auth.backends.ModelBackend'),
)
-------------------------------------------
```

Once the preceding command has been executed, the console should look similar to the following screenshot:

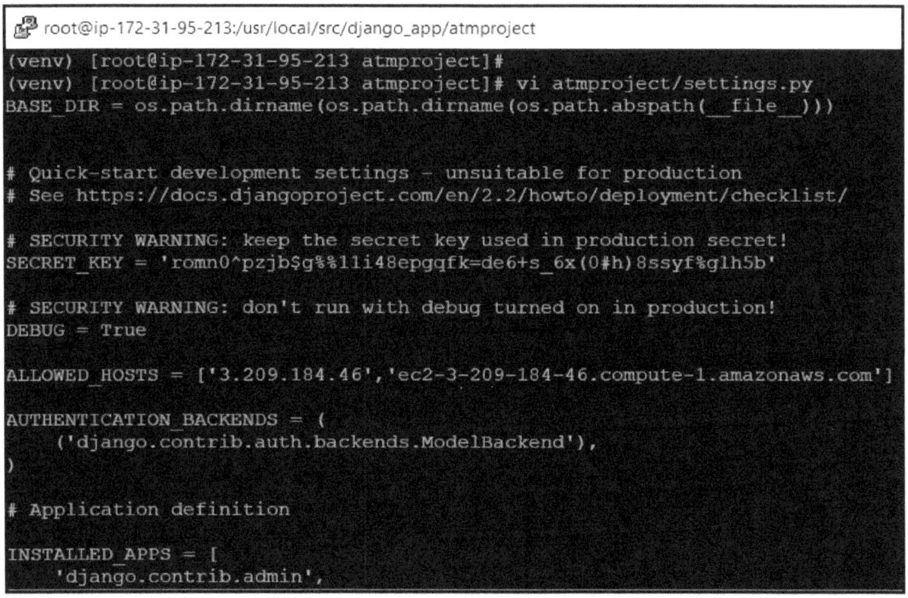

Figure 4.12 – The Django project settings

9. Now we can start our new Django project with the help of the following command:

```
(venv) [root@ip-172-31-95-213 atmproject]# python3.7 manage.py
runserver 0:8000
Watching for file changes with StatReloader
```

```
Performing system checks...
System check identified no issues (0 silenced).
June 10, 2019 - 06:29:24
Django version 2.2.2, using settings 'atmproject.settings'
Starting development server at http://0:8000/
Quit the server with CONTROL-C.
```

10. Next, open the
    `http://ec2-3-209-184-46.compute-1.amazonaws.com:8000/` link inside
    the browser. A successful message should appear in your browser, for
    instance, **The install worked successfully! Congratulations!:**

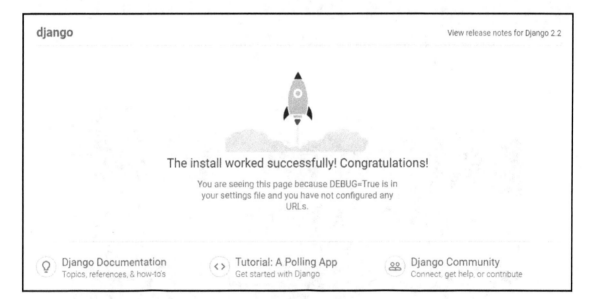

Figure 4.13 – Starting our Django project

11. We can also view the Django admin site with the help of the
    `http://ec2-3-209-184-46.compute-1.amazonaws.com:8000/admin` link.
    Use this link to show the Django admin site, as follows:

Figure 4.14 – The Django admin web

12. Use the Django superuser to log in with the following
   credentials: **Username**: admin and **Password**: djangoapp. This is shown in the
   following screenshot:

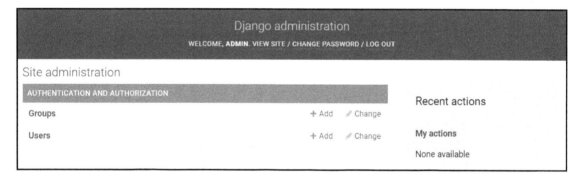

Figure 4.15 – Django administration

From the preceding general admin interface, we need a method to capture the application
data inside the admin web. In Django, this can be created using database models.

Database models in Django

After signing in to the Django admin, we have not yet loaded our ATM location data from the PostgreSQL RDS into our Django web. Therefore, we have to create our database model by performing the following steps:

1. The first step is to edit the `models.py` file; let's do it according to the following code block:

```
(venv) [root@ip-172-31-95-213 atmproject]# vi atmapp/models.py
------------------------------------------
from django.db import models
# Create your models here.
class ATMlocations(models.Model):
        ID = models.AutoField(primary_key=True)
        BankName = models.CharField(max_length=60)
        Address = models.CharField(max_length=50)
        County = models.CharField(max_length=15)
        City = models.CharField(max_length=15)
        State = models.CharField(max_length=2)
        ZipCode = models.IntegerField()
        class Meta:
                db_table = 'ATM locations'
                verbose_name = 'ATM location'
                verbose_name_plural = 'ATM locations'
------------------------------------------
```

After the preceding command has been executed, our console will appear as follows:

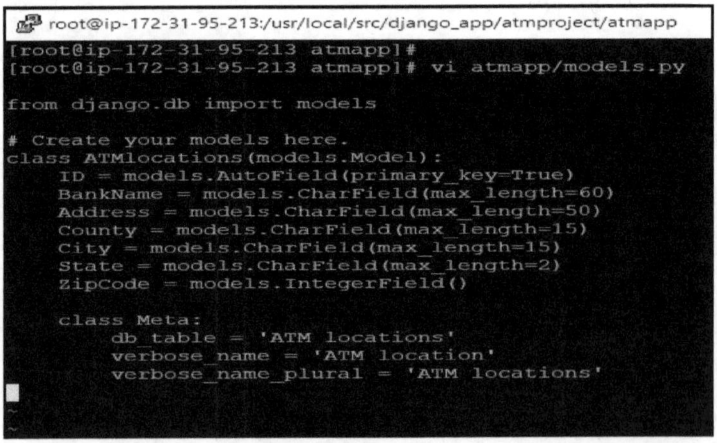

Figure 4.16 – The database model

2. The next step is to check for the app name inside the `apps.py` file with the help of the following command:

```
(venv) [root@ip-172-31-95-213 atmproject]# cat atmapp/apps.py
-----------------------------------------
from django.apps import AppConfig

class AtmappConfig(AppConfig):
        name = 'atmapp'
-----------------------------------------
```

3. In order to include the app within our project, we need to add a reference to its configuration class in the settings. This is `atmapp.apps.AtmappConfig`, as shown in the following screenshot:

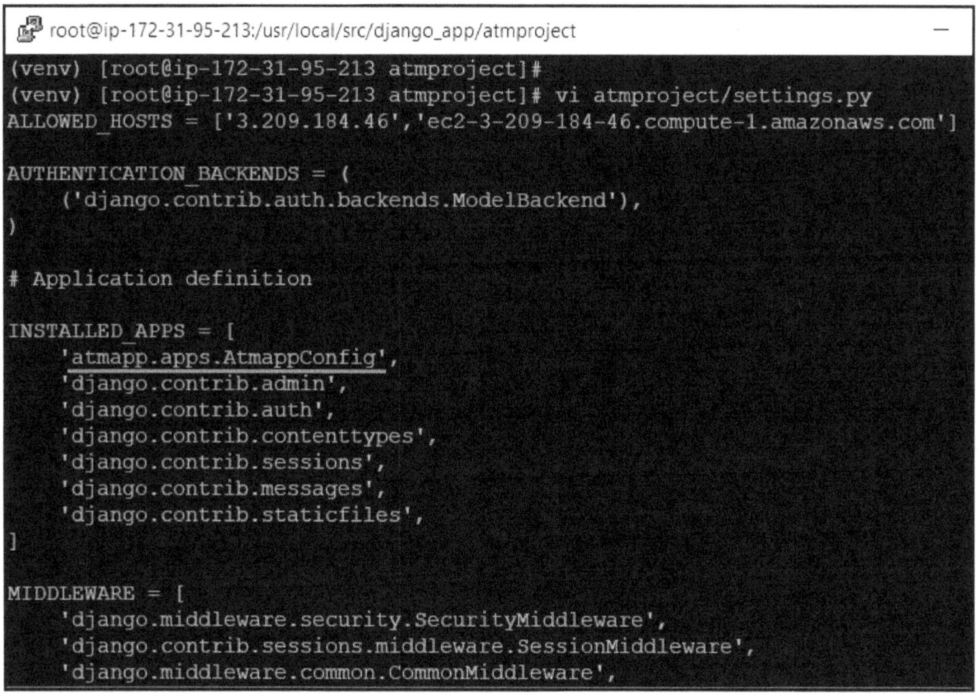

```
root@ip-172-31-95-213:/usr/local/src/django_app/atmproject                    —
(venv) [root@ip-172-31-95-213 atmproject]#
(venv) [root@ip-172-31-95-213 atmproject]# vi atmproject/settings.py
ALLOWED_HOSTS = ['3.209.184.46','ec2-3-209-184-46.compute-1.amazonaws.com']

AUTHENTICATION_BACKENDS = (
    ('django.contrib.auth.backends.ModelBackend'),
)

# Application definition

INSTALLED_APPS = [
    'atmapp.apps.AtmappConfig',
    'django.contrib.admin',
    'django.contrib.auth',
    'django.contrib.contenttypes',
    'django.contrib.sessions',
    'django.contrib.messages',
    'django.contrib.staticfiles',
]

MIDDLEWARE = [
    'django.middleware.security.SecurityMiddleware',
    'django.contrib.sessions.middleware.SessionMiddleware',
    'django.middleware.common.CommonMiddleware',
```

Figure 4.17 – Including an app inside the Django project

Of course, from the preceding database model, Django needs to be able to access the correct underlying table in the PostgreSQL RDS. In order to link the database model to the underlying table, you can try to migrate your database. We will demonstrate this in the next section.

Migrating the database

We can now update the new ATM locations model inside the Django app, as follows:

1. We will write the following `makemigrations` command, which is used to create the ATM locations model:

   ```
   (venv) [root@ip-172-31-95-213 atmproject]# python3.7 manage.py
   makemigrations atmapp
   Migrations for 'atmapp':
    atmapp/migrations/0001_initial.py
    - Create model ATMlocations
   ```

2. We will generate the SQL of `ATMlocations` using the `sqlmigrate` command:

   ```
   (venv) [root@ip-172-31-95-213 atmproject]# python3.7 manage.py
   sqlmigrate atmapp 0001
    BEGIN;
    --
    -- Create model ATMlocations
    --
    CREATE TABLE "ATM locations" ("ID" serial NOT NULL PRIMARY KEY,
   "BankName" varchar(60) NOT NULL, "Address" varchar(50) NOT NULL,
   "County" varchar(15) NOT NULL, "City" varchar(15) NOT NULL, "State"
   varchar(2) NOT NULL, "ZipCode" integer NOT NULL);
    COMMIT;
   ```

3. We will apply the migration to the PostgreSQL RDS, as shown in the following code block. However, since the ATM locations table already exists, the migrations will not be applied:

   ```
   (venv) [root@ip-172-31-95-213 atmproject]# python3.7 manage.py
   migrate
    Operations to perform:
            Apply all migrations: admin, atmapp, auth, contenttypes,
   sessions
    Running migrations:
            No migrations to apply.
   ```

   If the RDS does not contain the ATM locations table already, then this table will be created and the output will be as follows:

   ```
   Running migrations:
            Applying atmapp.0001_initial... OK
   ```

The database model is now ready to be registered inside the ATM app. We will do this in the next section. After that, we can try to make changes to our ATM locations' data using Django web.

Understanding the Django user interface – admin, views, templates, and URLs

A view inside a Django application generally performs a specific function and renders the results to a specific template. So, each view is a simple Python function or `class` method. Django will select a suitable view by examining the requested URL. Django URL patterns are very elegant; they are the general form of a URL. In order to select a view from the URL, Django uses a set of patterns called **URLconfs**. A URLconf will map URL patterns to their corresponding views.

Making the atmapp modifiable inside the admin

We need to tell the admin that the ATMlocations objects have an admin interface:

1. To do this, open the atmapp/admin.py file and edit it to look like the following code block:

   ```
   [root@ip-172-31-95-213 atmproject]# vi atmapp/admin.py
   ----------------------------------------
   from django.contrib import admin
   # Register your models here.
   from .models import ATMlocations
   admin.site.register(ATMlocations)
   ----------------------------------------
   ```

2. Then, we can launch the Django server with the help of the following command:

   ```
   (venv) [root@ip-172-31-95-213 atmproject]# python3.7 manage.py
   runserver 0:8000
   ```

3. After that, we will log in as the admin at
   `http://ec2-3-209-184-46.compute-1.amazonaws.com:8000/admin`.
   Using the `admin` and `djangoapp` login credentials, we will be able to view our
   captured ATM locations data from the PostgreSQL RDS, as shown in the
   following screenshot:

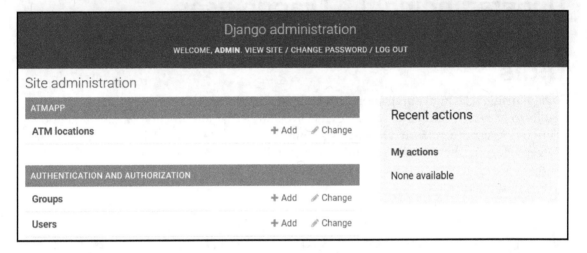

Figure 4.18 – Registering the database model inside a Django app

In the next section, we will discover which operations are supported by Django in the ATM
locations file.

Exploring the free admin functionality

The Django administration provides support for us to perform viewing, updating, adding
new, and deleting ATM location records. To carry out these tasks, perform the following
steps:

1. Click onto the ATM locations link; it will open 654 data objects, as shown in the
   following screenshot:

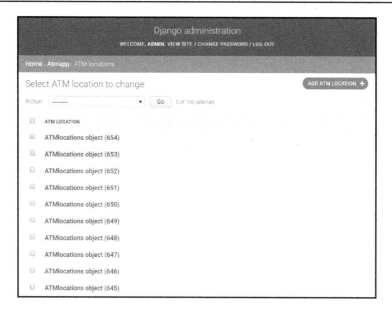

Figure 4.19 – Using the admin interface

2. Next, click on any of the data objects. Then, the details of its ATM will appear, as shown in the following screenshot:

Figure 4.20 – Viewing an ATM location

3. Now we will modify **BankName** to `Citibank NA` and then click on the **SAVE** button. We will get a confirmation message after we have saved the new bank name, as shown in the following screenshot:

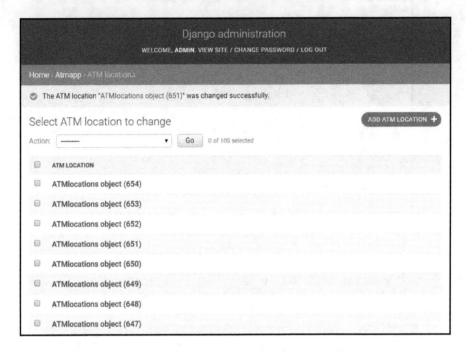

Figure 4.21 – Changing an ATM location

4. In the preceding screen of ATMapp (*Figure 4.20*), if you click on the **HISTORY** button, you can view the history of all the actions that have happened to that ATM, as shown in the following screenshot:

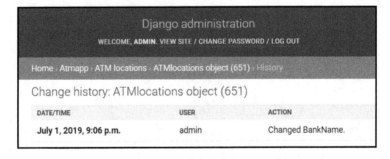

Figure 4.22 – Data object history

So far, you have performed all of the steps in our Django project of ATMs. Moving forward, you can apply the same steps to create your own Django projects.

Summary

In this chapter, we demonstrated a step-by-step process of how to implement the ATM banking PostgreSQL RDS inside a Django app. You learned how to set up Django within a virtual environment. We used the `postgresql_psycopg2` engine for Django to connect to our PostgreSQL RDS, and then we learned how to add a new database model. We activated the database model and managed it by using the built-in admin tool. Finally, we also learned how to create views, templates, and URLs.

In the next chapter, we will learn how to create geospatial functions using PostGIS for our PostgreSQL ATM database.

5

Creating a Geospatial Database Using PostGIS and PostgreSQL

This chapter will introduce you to PostGIS, a spatial extension for PostgreSQL that allows the user to store and spatially query geographic data. PostGIS is free and open source software. We will install the PostGIS extension for PostgreSQL, and then load spatial data into a database. Finally, after creating a spatial database, we will execute a few spatial queries that demonstrate some of its functionality.

Through the project demonstrated in this chapter, you will learn how to display a geographic map of a city in our banking PostgreSQL 12 RDS from AWS. Through PostGIS and QGIS, on the city map, all of the ATM locations will be marked and linked to an ATM network.

The following topics will be covered in this chapter:

- Installing PostGIS for RDS on AWS
- Importing spatial data files into PostgreSQL
- Setting up QGIS
- Loading spatial data using QGIS
- Executing PostGIS queries

Technical requirements

This chapter will take developers around 10-12 hours of work to develop a PostGIS application.

The code files for this chapter are available at `https://github.com/PacktPublishing/Developing-Modern-Database-Applications-with-PostgreSQL/tree/master/Chapter05`.

Installing PostGIS for RDS on AWS

On AWS, PostGIS installation is done through SQL statements because RDS does not provide support for PostGIS installation by source. Therefore, it is quite simple to create the extensions required by PostGIS using a few `CREATE EXTENSION` statements in our RDS. The steps are as follows:

1. Use pgAdmin to connect to our ATM RDS on AWS and select the ATM database to install PostGIS.
2. Then navigate to the top menu bar Tools ⇒ Query Tool, and then execute the below SQL statements by pressing the Execute/Refresh icon ▶ (or pressing the F5 key) on the toolbar

   ```
   CREATE EXTENSION postgis;
   CREATE EXTENSION postgis_topology;
   CREATE EXTENSION fuzzystrmatch;
   CREATE EXTENSION postgis_tiger_geocoder;
   CREATE EXTENSION address_standardizer;
   ```

   The following is a screenshot of the output after the first query is executed:

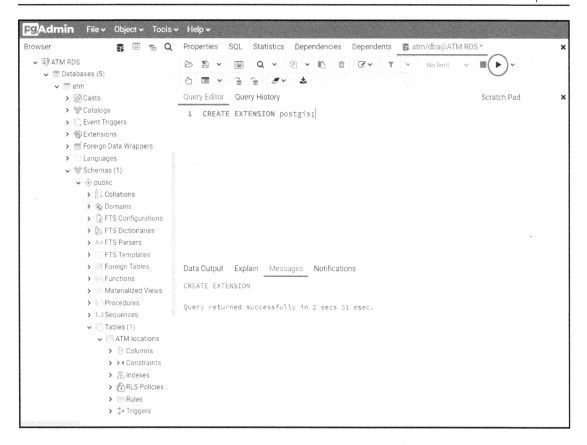

Figure 5.1 – Installing PostGIS for RDS on AWS

`postgis_topology` (PostGIS Topology) allows topological vector data to be stored in a PostGIS database.

`fuzzystrmatch` allows functions to determine string similarities and the distance between strings.

`postgis_tiger_geocoder` utilizes US Census data for geocoding that deals with US data; the program home page is at `https://www.census.gov/programs-surveys/geography.html`.

The Tiger extension is for geocoding any address in the US into corresponding latitude and longitude coordinates. The **Topologically Integrated Geographic Encoding and Referencing** (**TIGER**) system was released by the US Census Bureau. The census data part that this extension uses can be found at `https://www2.census.gov/geo/tiger/TIGER2019/`.

You can open all of the data structures from the preceding link in the browser as follows:

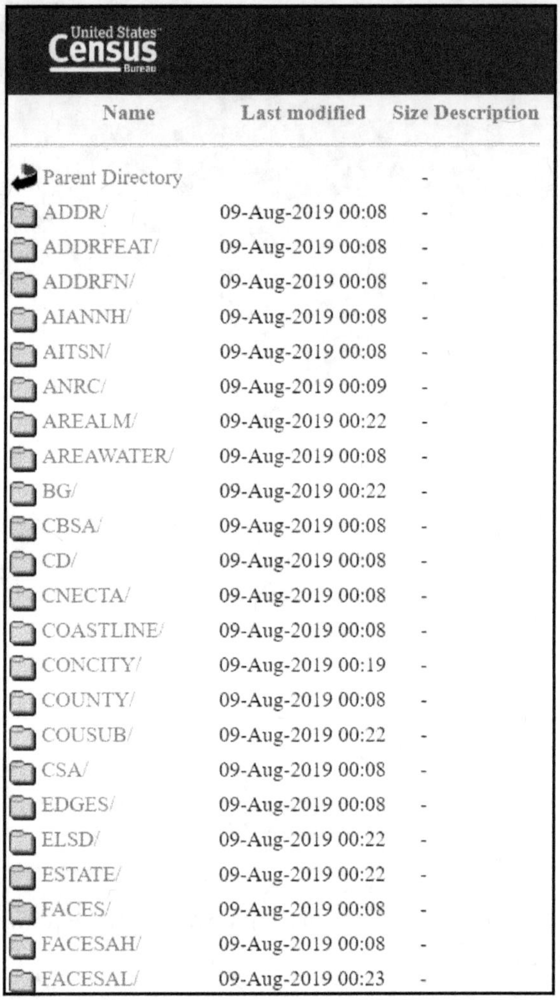
United States Census Bureau
Name Last modified Size Description
Parent Directory -
ADDR/ 09-Aug-2019 00:08 -
ADDRFEAT/ 09-Aug-2019 00:08 -
ADDRFN/ 09-Aug-2019 00:08 -
AIANNH/ 09-Aug-2019 00:08 -
AITSN/ 09-Aug-2019 00:08 -
ANRC/ 09-Aug-2019 00:09 -
AREALM/ 09-Aug-2019 00:22 -
AREAWATER/ 09-Aug-2019 00:08 -
BG/ 09-Aug-2019 00:22 -
CBSA/ 09-Aug-2019 00:08 -
CD/ 09-Aug-2019 00:08 -
CNECTA/ 09-Aug-2019 00:08 -
COASTLINE/ 09-Aug-2019 00:08 -
CONCITY/ 09-Aug-2019 00:19 -
COUNTY/ 09-Aug-2019 00:08 -
COUSUB/ 09-Aug-2019 00:22 -
CSA/ 09-Aug-2019 00:08 -
EDGES/ 09-Aug-2019 00:08 -
ELSD/ 09-Aug-2019 00:22 -
ESTATE/ 09-Aug-2019 00:22 -
FACES/ 09-Aug-2019 00:08 -
FACESAH/ 09-Aug-2019 00:08 -
FACESAL/ 09-Aug-2019 00:23 -

Figure 5.2 – US Census data for geocoding

`address_standardizer` is a single-line address parser that takes an input address and normalizes it based on a set of rules.

3. Now you should see `postgis`,
   `postgis_topology`, `postgis_tiger_geocoder`, and more as available
   extensions in **PgAdmin | Extensions**. Execute the following query to verify
   whether the extensions are installed successfully:

   ```
   SELECT name, default_version,installed_version
   FROM pg_available_extensions WHERE name LIKE 'postgis%' or name
   LIKE 'address%';
   ```

As shown in the following screenshot, for PostgreSQL 12, the PostGIS versions are all
3.0.0 together:

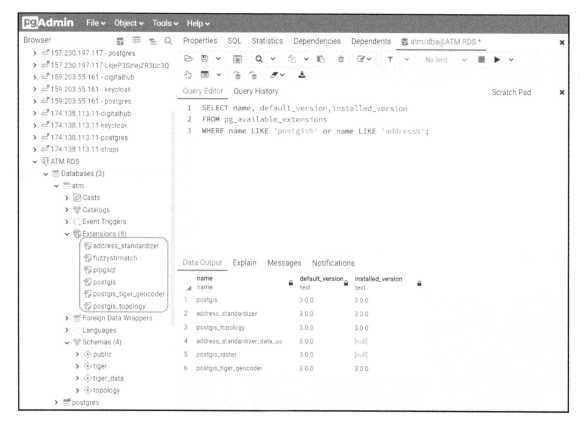

Figure 5.3 – Verifying PostGIS installation

When you can reach the preceding result, which shows clearly six extensions with versions,
it means that your PostGIS installation has been done and you are going to have PostGIS
working smoothly without any issues for the next steps.

Importing spatial data files into PostgreSQL

For each of the ATM locations that we currently store inside our ATM database, we will have to supply the geographical data, including longitude and latitude data. After providing these geographical points for PostGIS, the extension will be able to show a world map with location points on it. The steps for this are as follows:

1. Download the `zipcoordinates.sql` file from GitHub at `https://github.com/lequanha/PostgreSQL-12-Development-and-Administration-Projects/blob/master/Chapter 5/zipcoordinates.sql`. On your browser, you will see that the script includes a `CREATE TABLE` statement and an `INSERT` statement to create and to populate a new table named `Zip coordinates`:

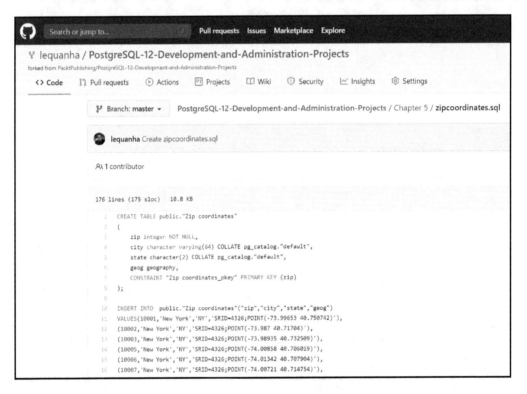

Figure 5.4 – GitHub file for spatial data

2. Next, execute this SQL file inside pgAdmin to create the new `Zip coordinates` database table inside the PostgreSQL RDS, as shown in the following screenshot:

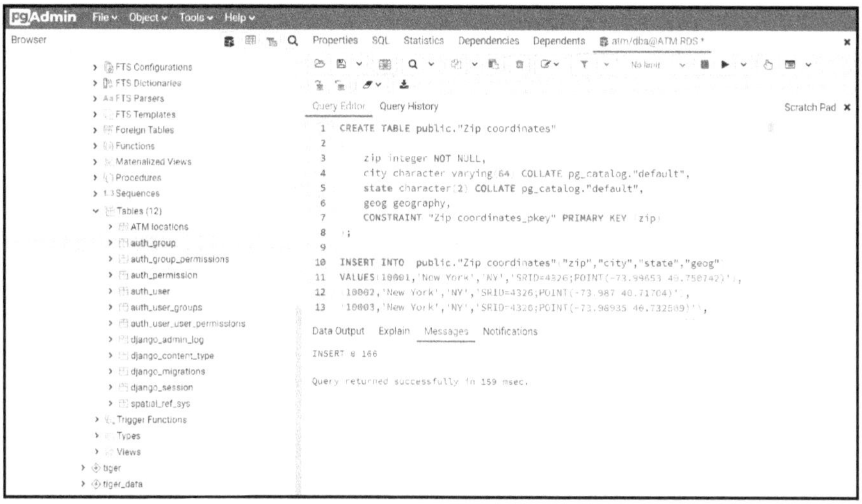

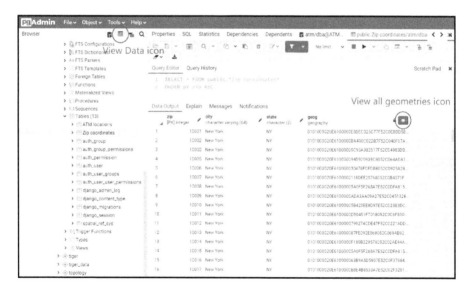

Figure 5.5 – Creating the Zip coordinates table

3. Right-click on **Tables** in the left panel and click on **Refresh** and you will see that the new `Zip coordinates` table will pop up. After that, you can click on the **View Data** icon on the toolbar to view all the zip locations, as shown in the following screenshot:

Figure 5.6 – Zip codes' geography data

4. Click on the **View all geometries** icon on the left of the `geog` column header (*Figure 5.6*) to view the map, and then use the two **Zoom in / Zoom out** icons to enlarge the map so that you can see the actual locations corresponding to the zip coordinates for New York City, as shown in the following screenshot:

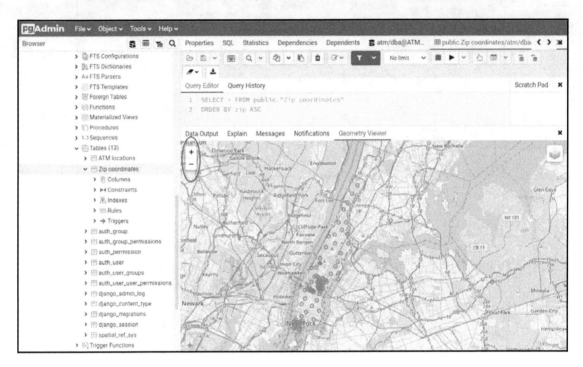

Figure 5.7 – Real coordinates of the zip codes of New York

5. Combine the `Zip coordinates` table with the ATM locations table into a new view called `ATM coordinates` in which the real coordinates of the ATM machines from New York City can be linked. The following is the view definition:

```
CREATE VIEW "ATM coordinates" AS
SELECT al.*, zc."geog"
FROM "ATM locations" al
INNER JOIN "Zip coordinates" zc
ON al."ZipCode" = zc."zip";
```

The following screenshot shows the view execution on pgAdmin:

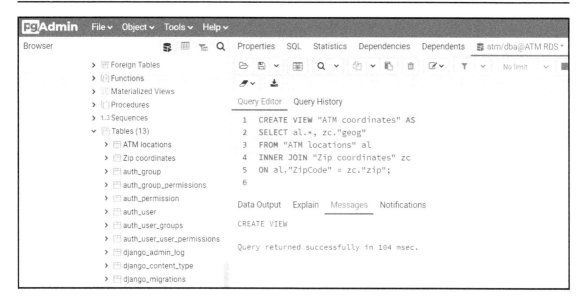

Figure 5.8 – Creation of the new ATM coordinates view

6. Expand the **Views** entry on the left panel to select the new ATM coordinates view, and then click the **View Data** icon on the toolbar to view the ATM locations with the real coordinates, as shown in the following screenshot:

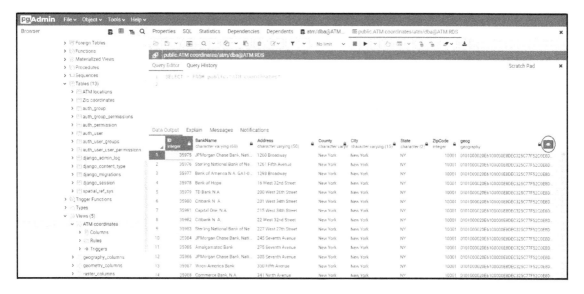

Figure 5.9 – ATM coordinates data

7. Click on the **View all geometries** icon on the left of the `geog` column header (*Figure 5.9*) to view the map, and then use the two **Zoom in/Zoom out** icons to enlarge the map so that the real points for the ATM machines of New York City can be seen, as shown in the following screenshot:

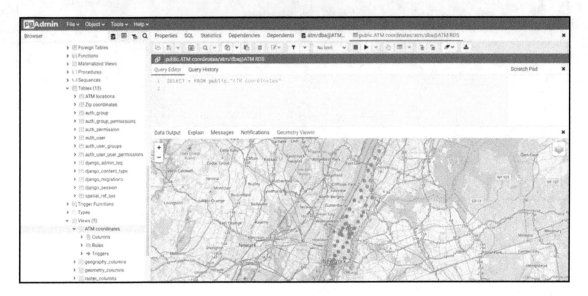

Figure 5.10 – The ATM network of New York City

Geometry Viewer in PostGIS is very convenient and easy to use. Because this viewer tool employs OpenStreetMap, our data points have been set to SRID = 4326 in our later spatial queries.

Setting up QGIS

QGIS, previously known as **Quantum GIS**, is a free open source tool for viewing, editing, and analyzing geospatial data. To handle spatial data, follow these steps to establish a connection between PostGIS and QGIS:

1. Choose the below **Amazon Machine Image (AMI)**: Microsoft Windows Server 2019 Base - `ami-077f1edd46ddb3129`. This AMI can be seen at the top in the following screenshot:

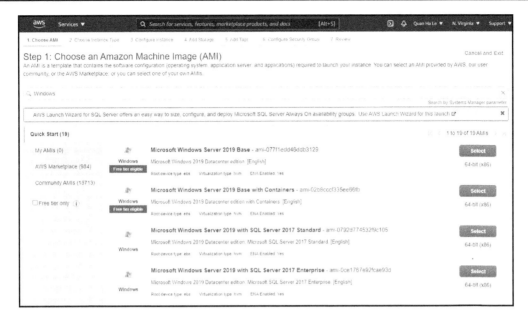

Figure 5.11 – AMI for Microsoft Windows Server

2. Launch a new EC2 instance from the AMI, as shown in the following screenshot:

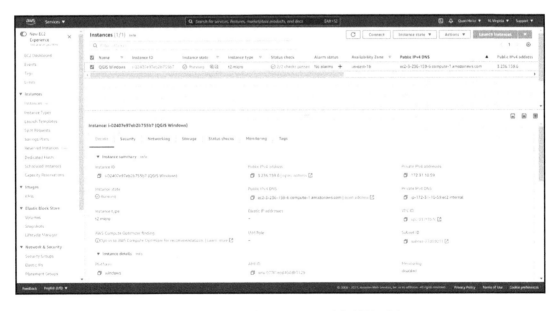

Figure 5.12 – The new Microsoft Windows Server is ready for QGIS installation

3. Open the new EC2 instance using remote desktop connection, as shown in the following screenshot, and then download QGIS from `https://qgis.org/en/site/forusers/download.html`. The QGIS installation package is `osgeo4w-setup-x86_64.exe`:

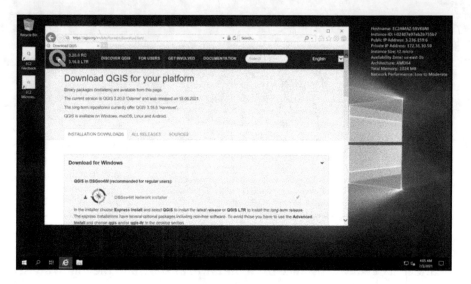

Figure 5.13 – Download QGIS

4. Select the **Express Desktop Install** type for QGIS, as shown in the following screenshot:

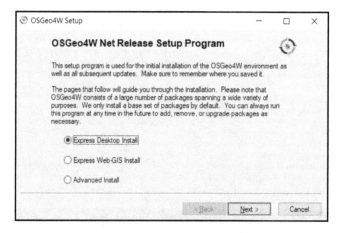

Figure 5.14 – Express Desktop Install

The following screenshot shows the installation progress:

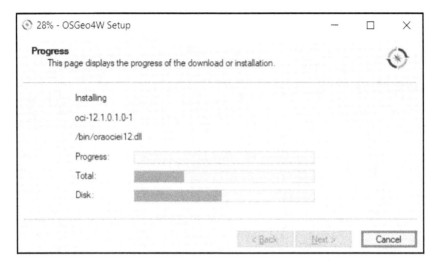

Figure 5.15 – QGIS installation progress

5. Wait until the installation is complete, as shown in the following screenshot, and then click on **Finish**:

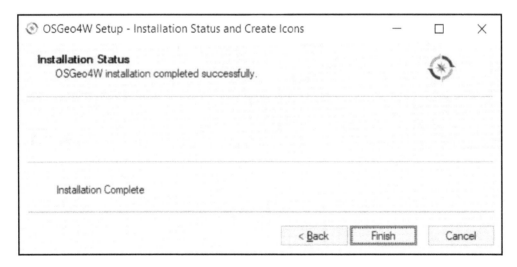

Figure 5.16 – QGIS installation complete

Now we can see that the QGIS software was installed from the Windows EC2 instance, as shown in the following screenshot. After that, you will need to click on **QGIS Desktop 3.8.1** to launch it:

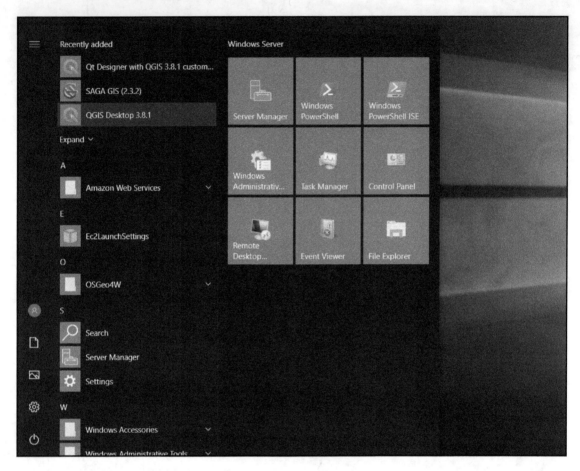

Figure 5.17 – QGIS inside the EC2 instance

QGIS installation on Windows 10 is very simple and fast. We are now ready to connect our QGIS to the PostGIS database in the next section.

Loading spatial data using QGIS

QGIS is a very common open source tool for GIS analysts, geographers, civil engineers, and many other professionals for creating maps and editing, viewing, and analyzing geospatial data. PostGIS is open source for PostgreSQL databases only, whereas QGIS is a wider open source spatial tool for many other databases, such as MSSQL, Oracle, and DB2. Hence, many developers have chosen QGIS as their GIS tool. In the following steps, we will see how PostGIS can be integrated into QGIS by setting a connection between them:

1. Right-click on **PostGIS** in the list under the **Browser** panel, as shown in the following screenshot, and select **New Connection**:

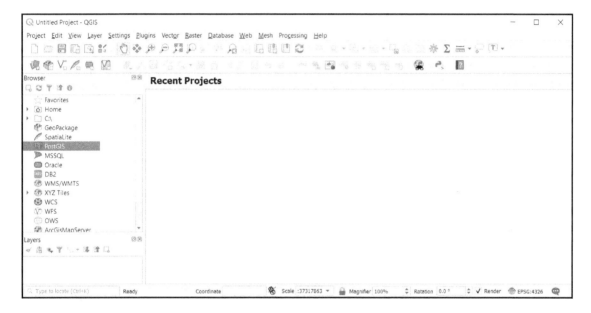

Figure 5.18 – QGIS main window

2. Fill in the RDS information, as shown in the following screenshot, click on the **Basic** tab for authentication, and then enter the following PostgreSQL database credentials:

- **User name**: dba
- **Password**: bookdemo

Figure 5.19 – Connecting QGIS to PostGIS

3. Now test your connection and then click **OK**. By using the zoom icon, you should get your ATMs on the map, as shown in the following screenshot:

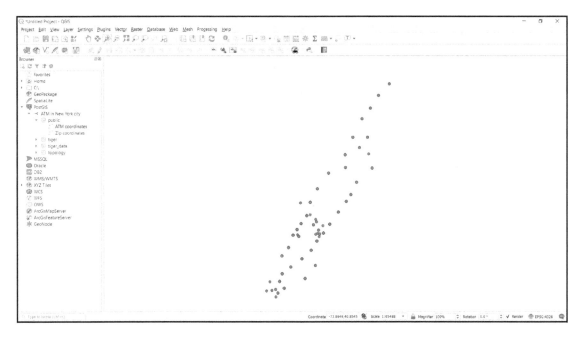

Figure 5.20 – The ATM networks shown by QGIS

Now, your QGIS is connected to PostGIS spatial data successfully. You can explore more functions of QGIS by referring to the main site of QGIS at `https://www.qgis.org/en/site/`. We will show you how to write a spatial `SELECT` statement using PostGIS in the next section.

Executing PostGIS queries

In this section, we will run some spatial queries in pgAdmin to demonstrate some PostGIS functionality. We will go back to pgAdmin and get into our ATM database to try to run a couple of sample queries. The first query will list ATM locations by their proximity to the Brooklyn Bridge, and the second query will capture all of the ATM locations within 1 kilometer of another specific place, Times Square.

Ordering ATM locations by distance from the Brooklyn Bridge

The steps for this are as follows:

1. We will use latitude `40.709677` and longitude `-74.00365` for the Brooklyn Bridge in the following SQL statement:

```
SELECT atm."BankName", atm."Address"
FROM "ATM coordinates" atm
ORDER BY geog <-> ST_SetSRID(ST_MakePoint(-74.00365, 40.709677),
4326);
```

The result of the preceding query is shown in the following screenshot:

Figure 5.21 – The ATM machines ordered by proximity to the Brooklyn Bridge

2. You can change the location from the Brooklyn Bridge to any other location to test your `SELECT` query; for instance, you could use New York City Hall (the latitude and longitude coordinates are `40.712772, -74.006058`), the Rockefeller Center (latitude `40.758740`, longitude `-73.978674`), and the Metropolitan Museum of Art (latitude `40.778965`, longitude `-73.962311`).

Finding ATM locations within 1 kilometer of Times Square

The steps for this are as follows:

1. The second query is a common issue for every pedestrian. Suppose that you are visiting Times Square in New York City and you suddenly need to find the nearest ATM. We will use latitude `40.755101` and longitude `-73.99337` for Times Square and use `1000` meters to form the following SQL statement:

```
SELECT atm."BankName", atm."Address"
FROM "ATM coordinates" atm
WHERE ST_DWithin(geog, ST_SetSRID(ST_MakePoint(-73.99337,
40.755101), 4326), 1000);
```

The result of the preceding query is shown in the following screenshot:

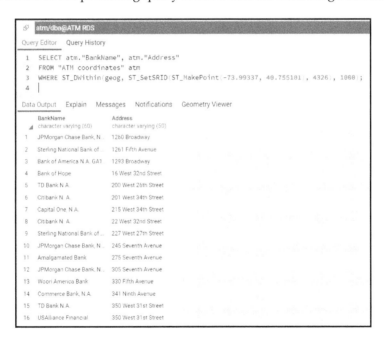

Figure 5.22 – The ATMs nearest Times Square

We can also practice around with this query by changing Time Square to other locations into the preceding query such as New York City Hall, the Rockefeller Center, the Metropolitan Museum of Art, and the Brooklyn Bridge, and so on.

Summary

This chapter has introduced you to GIS and spatial data in PostgreSQL via a step-by-step project using the PostGIS extension. In this project, you first learned how to install PostGIS for RDS on AWS. Then, you learned how to declare and import spatial data files into PostGIS and how to show spatial data with the Geometry Viewer feature of PostGIS. After that, we also showed you another GIS open source variant named QGIS so that you can broaden your ideas about GIS, since QGIS is a wider tool that can also be applied to many other databases, such as Oracle, DB2, and MSSQL databases. Finally, we concluded the chapter with a couple of sample spatial queries for ATM locations in New York City.

In the next chapter, we will study PostgREST, a RESTful API for PostgreSQL databases.

6
Managing Banking Transactions using PostgREST

In this chapter, we will learn how to develop PostgREST. It is a standalone web server that automatically generates usable RESTful APIs from any PostgreSQL databases directly. The API endpoints and operations are implemented by the permissions and structural constraints of the PostgreSQL database. PostgREST serves as an automatic solution to the manual **CRUD** (**Create, Read, Update, and Delete**) querying.

The project in this chapter will use PostgREST and Docker to set up a RESTful API for banking transactions to send GET, POST, PUT, and DELETE `curl` requests into the PostgreSQL 12 **Relational Database Service** (**RDS**) referred to in Chapter 2, *Setting Up a PostgreSQL RDS for ATM Machines*, from **Amazon Web Services** (**AWS**) to create, retrieve, update, and delete ATM machine locations within a typical city.

The following topics will be covered in the chapter:

- Introduction to PostgREST
- Creating a PostgREST API schema on an RDS
- Executing PostgREST
- Adding a trusted user
- Creating a PostgREST token
- PostgREST administration
- PostgREST on TimescaleDB

Technical requirements

This chapter will take developers around 10-12 hours of working to develop PostgREST.

The figures and code files of this chapter are available at the following link: `https://github.com/PacktPublishing/Developing-Modern-Database-Applications-with-PostgreSQL/tree/master/Chapter06`.

Introduction to PostgREST

In this section, we are going to practice Docker installation and will learn how to start Docker as a service. We will download the PostgREST image from Docker to start the first launching of PostgREST.

Using Docker

Docker is a tool that uses containers to package an application with all of its libraries and other dependencies and ship it all out as one package:

1. We will PuTTY into our `ec2` instance with the built-in user for our current AMI, that is, the `centos` user, with the help of the following command:

   ```
   [centos@ip-172-31-95-213 ~]$ sudo su
   [root@ip-172-31-95-213 centos]# cd /usr/local/src/
   ```

2. You must check whether Docker is installed. If it is not installed, then you can get it installed with the help of the following command:

   ```
   [root@ip-172-31-95-213 src]# yum install -y yum-utils device-
   mapper-persistent-data lvm2
   [root@ip-172-31-95-213 src]# yum-config-manager --add-repo
   https://download.docker.com/linux/centos/docker-ce.repo
   [root@ip-172-31-95-213 src]# yum install docker-ce
   ```

3. During the installation, type `y` (`=Yes`) on the terminal if you are asked any questions, as shown in *Figure 6.1*:

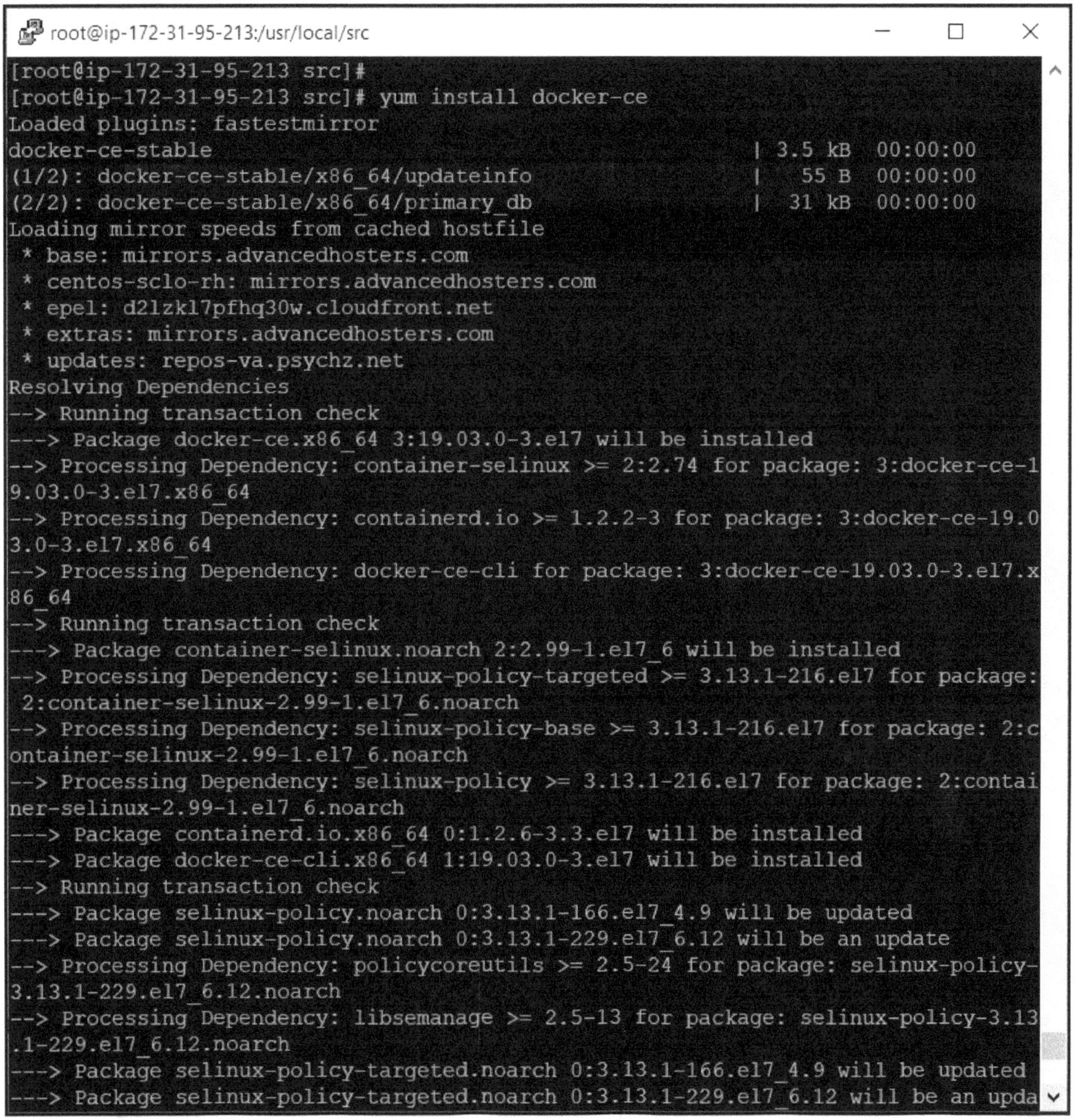

```
root@ip-172-31-95-213:/usr/local/src                              —    □    ×
[root@ip-172-31-95-213 src]#
[root@ip-172-31-95-213 src]# yum install docker-ce
Loaded plugins: fastestmirror
docker-ce-stable                                     | 3.5 kB  00:00:00
(1/2): docker-ce-stable/x86_64/updateinfo            |  55 B   00:00:00
(2/2): docker-ce-stable/x86_64/primary_db            |  31 kB  00:00:00
Loading mirror speeds from cached hostfile
 * base: mirrors.advancedhosters.com
 * centos-sclo-rh: mirrors.advancedhosters.com
 * epel: d2lzkl7pfhq30w.cloudfront.net
 * extras: mirrors.advancedhosters.com
 * updates: repos-va.psychz.net
Resolving Dependencies
--> Running transaction check
---> Package docker-ce.x86_64 3:19.03.0-3.el7 will be installed
--> Processing Dependency: container-selinux >= 2:2.74 for package: 3:docker-ce-1
9.03.0-3.el7.x86_64
--> Processing Dependency: containerd.io >= 1.2.2-3 for package: 3:docker-ce-19.0
3.0-3.el7.x86_64
--> Processing Dependency: docker-ce-cli for package: 3:docker-ce-19.03.0-3.el7.x
86_64
--> Running transaction check
---> Package container-selinux.noarch 2:2.99-1.el7_6 will be installed
--> Processing Dependency: selinux-policy-targeted >= 3.13.1-216.el7 for package:
 2:container-selinux-2.99-1.el7_6.noarch
--> Processing Dependency: selinux-policy-base >= 3.13.1-216.el7 for package: 2:c
ontainer-selinux-2.99-1.el7_6.noarch
--> Processing Dependency: selinux-policy >= 3.13.1-216.el7 for package: 2:contai
ner-selinux-2.99-1.el7_6.noarch
---> Package containerd.io.x86_64 0:1.2.6-3.3.el7 will be installed
---> Package docker-ce-cli.x86_64 1:19.03.0-3.el7 will be installed
--> Running transaction check
---> Package selinux-policy.noarch 0:3.13.1-166.el7_4.9 will be updated
---> Package selinux-policy.noarch 0:3.13.1-229.el7_6.12 will be an update
--> Processing Dependency: policycoreutils >= 2.5-24 for package: selinux-policy-
3.13.1-229.el7_6.12.noarch
--> Processing Dependency: libsemanage >= 2.5-13 for package: selinux-policy-3.13
.1-229.el7_6.12.noarch
---> Package selinux-policy-targeted.noarch 0:3.13.1-166.el7_4.9 will be updated
---> Package selinux-policy-targeted.noarch 0:3.13.1-229.el7_6.12 will be an upda
```

Figure 6.1 – Installing Docker

4. Add your user to the `docker` group with the help of the following command:

```
[root@ip-172-31-95-213 src]# usermod -aG docker $(whoami)
```

5. Set Docker to start automatically at boot time by typing the following command:

```
[root@ip-172-31-95-213 src]# systemctl enable docker.service
```

6. Finally, start the Docker service by running the following command:

```
[root@ip-172-31-95-213 src]# systemctl start docker.service
```

In order to run Docker, you have to start the Docker service or the Docker daemon as we have just learned how to start it in the previous statement.

Installing standalone PostgREST

After we have started the Docker daemon, we can run Docker commands without daemon connection errors:

1. Let's pull and start the PostgREST image from Docker as shown here:

```
[root@ip-172-31-95-213 src]# docker run --name tutorial -p
5432:5432 -e POSTGRES_PASSWORD=mysecretpassword -d postgres
```

This will run the Docker instance as a daemon and expose port 5432 to the host system so that it looks like an ordinary PostgreSQL server to the rest of the system, as shown in *Figure 6.2*:

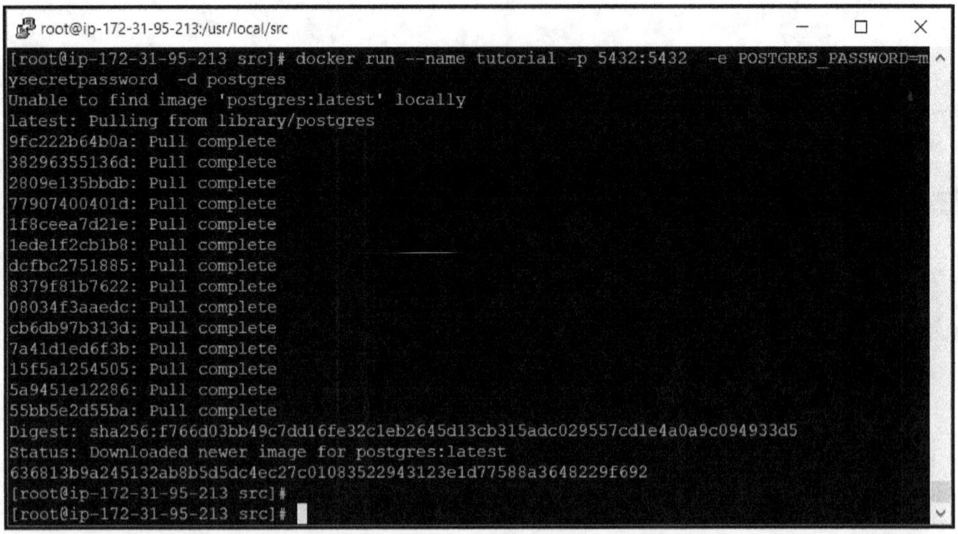

Figure 6.2 – Pulling a PostgREST image from Docker

2. Next, we will connect to the SQL console (`psql`) inside the container with the help of the following command:

```
[root@ip-172-31-95-213 src]# docker exec -it tutorial psql -U
postgres
psql (11.5 (Debian 11.5-1.pgdg90+1))
Type "help" for help.
postgres=#
```

Once the preceding command is executed, we will see the `psql` command prompt. You can start to create your database schema and tables here so that later on you will have the PostgREST API connecting to these local databases.

Standalone PostgREST can deploy the following two features:

- **TimescaleDB**: An open source database for time series data that provides automatic partitioning across time and space for PostgreSQL
- **PostgREST Test Suite**: A unit test library that provides a sample test PostgreSQL database with creation scripts to deploy a sample testing PostgREST service onto that test database, and finally supplying auto-testing scripts to check that your testing PostgREST service can have good memory usage and to send sample input/output to make sure that your testing PostgREST service works well

Next, we will switch away from standalone PostgREST as we will have to test PostgREST on AWS in order to deploy our banking application.

Creating a PostgREST API schema on an RDS (AWS)

We will install PostgREST on a CentOS `ec2` instance, and then we will set up PostgREST to create an API for our PostgreSQL version RDS on the AWS cloud:

1. Let's install PostgREST from the binary release by using the command shown in the following code block:

```
[root@ip-172-31-95-213 src]# yum install postgresql-libs
Please answer y (=Yes) when being asked,
[root@ip-172-31-95-213 src]# wget
https://github.com/PostgREST/postgrest/releases/download/v6.0.1/pos
tgrest-v6.0.1-centos7.tar.xz
[root@ip-172-31-95-213 src]# tar xfJ postgrest-v6.0.1-
centos7.tar.xz
```

```
[root@ip-172-31-95-213 src]# mv postgrest /usr/local/bin/
[root@ip-172-31-95-213 src]# rm postgrest-v6.0.1-centos7.tar.xz
```

2. We will now define a user role that has permission to access the RDS ATM database and to read records from the ATM locations table, as shown:

```
create role web_anon nologin;
grant web_anon to dba;

grant usage on schema public to web_anon;
grant select on public."ATM locations" to web_anon;
```

After executing the preceding query, we will get the following result as shown in *Figure 6.3*:

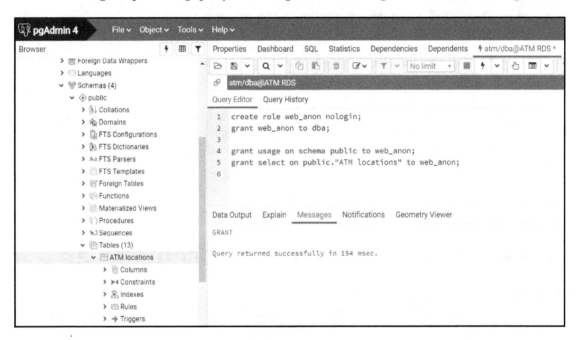

Figure 6.3 – Creating the PostgREST user role

For PostgREST to connect with our RDS on the AWS cloud, we should remember the name assigned for the new role. In the example demonstrated here, we have assigned the name web_anon to the new role.

Executing PostgREST

We are now putting together what we have set up until now, the database schema, the user role, and the PostgreSQL version 12 connection of the RDS, into PostgREST in the following steps:

1. We will create the PostgREST configuration file as shown and will name it `tutorial.conf`:

```
[root@ip-172-31-95-213 src]# mkdir postgrest
[root@ip-172-31-95-213 src]# cd postgrest
[root@ip-172-31-95-213 postgrest]# vi tutorial.conf
db-uri = "postgres://dba:bookdemo@atm.ck5074bwbilj.us-
east-1.rds.amazonaws.com/atm" db-schema = "public" db-anon-role =
"web_anon"
```

2. Now, we will call PostgREST to create an API on our RDS with the help of the following command:

```
[root@ip-172-31-95-213 postgrest]# postgrest tutorial.conf
```

PostgREST will start listening on port 3000 after it is connected to the RDS, hence the automatic API on the ATM database has been established. After you run the preceding command, your console will look as shown in the following screenshot:

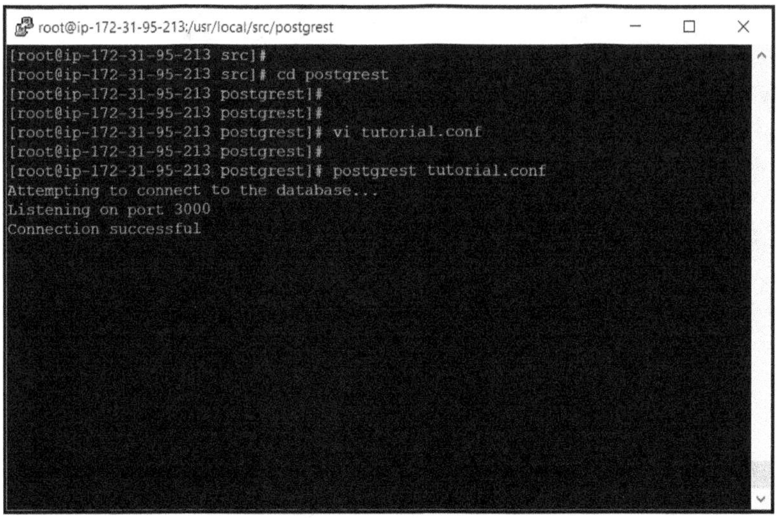

Figure 6.4 – PostgREST server

3. Open a new terminal, leaving the current terminal but continuing to keep PostgREST running inside that one so that we have two terminals at the same time. Try doing an HTTP request for the ATM locations as shown:

```
[root@ip-172-31-95-213 centos]# curl
http://localhost:3000/ATM%20locations
```

The space character is replaced by %20 inside the curl statement. Please observe the curl result in *Figure 6.5*:

Figure 6.5 – Getting all ATM locations

4. With the permissions allotted to the current role, anonymous requests have read-only access to the ATM locations table as shown in the following code block. If we try to add a new ATM, we will not be able to do so:

```
[root@ip-172-31-95-213 centos]# curl
http://localhost:3000/ATM%20locations -X POST -H "Content-Type:
application/json" -d '{"BankName":"Test Bank","Address":"99 Test
way","County":"New York","City":"New
York","State":"NY","ZipCode":10271}'
Response is 401 Unauthorized:
{"hint":null,"details":null,"code":"42501","message":"permission
denied for table ATM locations"}
```

Now we have established a basic API on the database by PostgREST. As you see, the POST request to insert a new ATM location has failed because web_anon was granted select privileges only.

Adding a trusted user

In order to perform more data manipulation on PostgREST, we will have to employ user roles with all privileges granted on the ATM locations table:

1. First, we will execute the following query using pgAdmin to create a new role:

```
create role atm_user nologin;
grant atm_user to dba;

grant usage on schema public to atm_user;
grant all on public."ATM locations" to atm_user;
grant usage, select on sequence public."ATM locations_ID_seq" to
atm_user;
```

On execution of the preceding script, the result will be as shown in *Figure 6.6*:

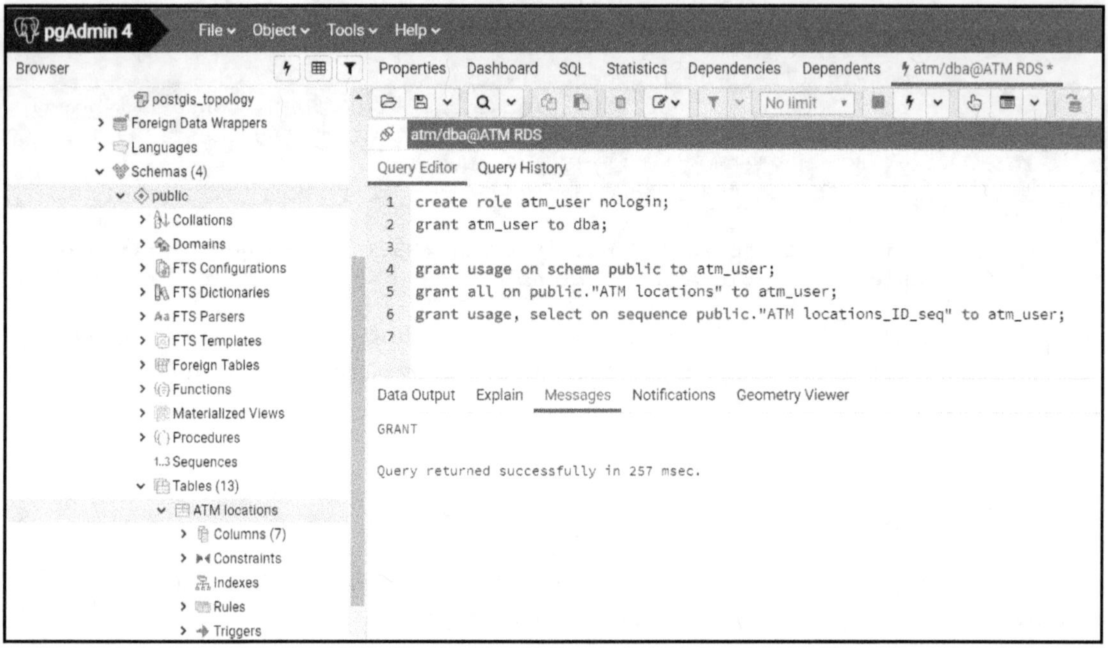

Figure 6.6 – Adding a trusted user

2. Next, we will set a password and provide it to PostgREST. You can use the following statements to generate a random password or you can think out a nice password as well; this password must be at least 32 characters long:

```
[root@ip-172-31-95-213 postgrest]# export LC_CTYPE=C
[root@ip-172-31-95-213 postgrest]# < /dev/urandom tr -dc A-Za-z0-9
| head -c32 DFZ49GQGubpzcSbt3t2uMIiBF6pU4PJ8
[root@ip-172-31-95-213 postgrest]#
```

/dev/urandom is a random character generator for Linux and while <
/dev/urandom generates a random string, the tr statement is extracting from
that random string the first 32 characters within A-Za-z0-9 by the head
statement.

3. We will add this generated password to the tutorial.conf file as shown:

```
jwt-secret = "<the password you made>"
```

On the console, the preceding command will look as shown in *Figure 6.7*:

Figure 6.7 – Making a secret

4. Because we still kept our previous terminal with the PostgREST server running inside, we will now restart it to load the updated configuration file (please press *Ctrl + C* to stop the server, then start PostgREST again) as shown in *Figure 6.8*:

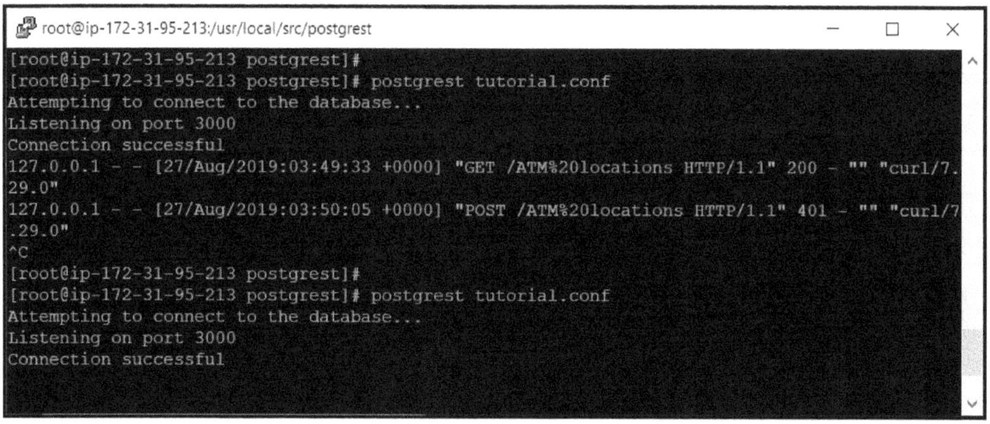

Figure 6.8 – Reloading the new configuration file

Now, whoever intends to use our PostgREST API will have to insert the correct password to issue insertion requests from the previously created user.

Creating a PostgREST token

In order to make efficient usage of the trusted user and the password of PostgREST, we will combine these two values into a token so that any API requests carrying that token will be accepted by PostgREST as the correct password and correct trusted user:

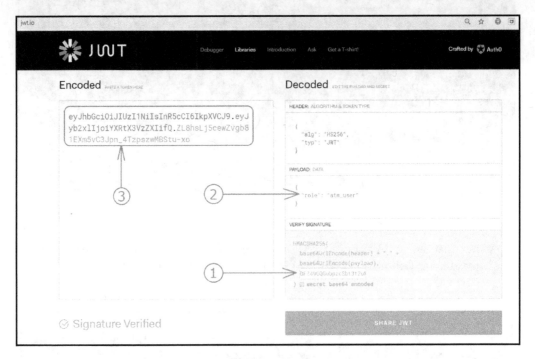

Figure 6.9 – Creating a token

We'll get started with creating the token using the following steps:

1. We will first open `https://jwt.io` on the browser and fill in the required details as shown:
   - **Password**: `DFZ49GQGubpzcSbt3t2uMIiBF6pU4PJ8`
   - **JSON**: `"role": "atm_user"`
   - **Copy the result**
     token: `eyJhbGciOiJIUzI1NiIsInR5cCI6IkpXVCJ9.eyJyb2xlIjoi YXRtX3VzZXIifQ.ZL8hsLj5cewZvgb81EXm5vC3Jpn_4TzpszwMBStu -xo`

   You can see the preceding values are entered in *Figure 6.9*.

2. We will now write the following command to be able to use the encoded token to add a new ATM location, on the other terminal (not the one that PostgREST is running inside):

```
[centos@ip-172-31-95-213 ~]$ export
TOKEN="eyJhbGciOiJIUzI1NiIsInR5cCI6IkpXVCJ9.eyJyb2xlIjoiYXRtX3VzZXI
ifQ.ZL8hsLj5cewZvgb81EXm5vC3Jpn_4TzpszwMBStu-xo"
[centos@ip-172-31-95-213 ~]$ curl
http://localhost:3000/ATM%20locations -X POST -H "Authorization:
Bearer $TOKEN" -H "Content-Type: application/json" -d
'{"BankName":"Test Bank","Address":"99 Test way","County":"New
York","City":"New York","State":"NY","ZipCode":10271}'
```

Because you authorize your POST request with the correct token as shown previously, when PostgREST receives your request, it assumes that your request connects by the correct password using atm_user, who has all the privileges on the ATM locations table. After the preceding command is executed, the console will have no errors, as shown in *Figure 6.10*:

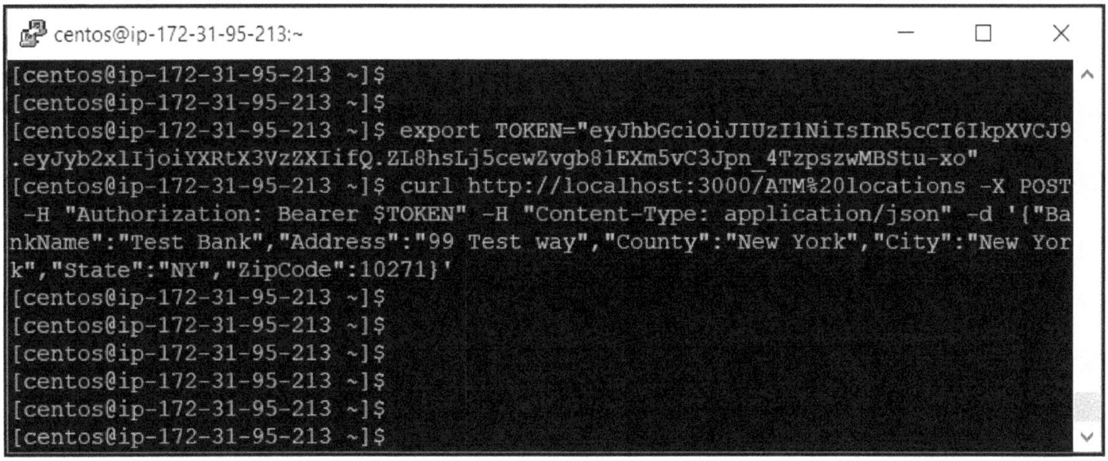

Figure 6.10 – POST an insertion by PostgREST

3. After that, we see the new test ATM location by pgAdmin as in *Figure 6.11*:

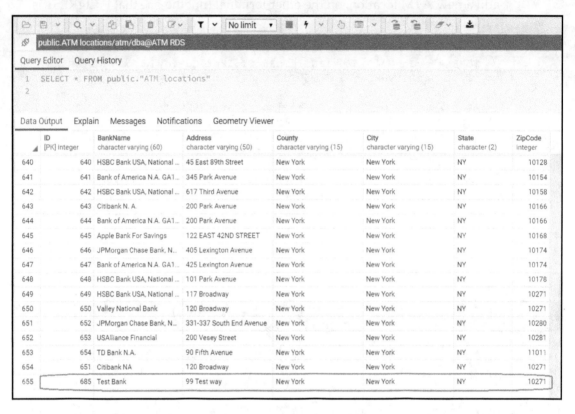

Figure 6.11 – The insertion has been successful

4. We create a token that will expire after 5 minutes instead of the forever alive token previously. At first, we calculate the epoch value of 5 minutes with the help of the following command:

```
select extract(epoch from now() + '5 minutes'::interval) ::
integer;
```

The result of the preceding `select` statement will be the epoch value as shown in *Figure 6.12*:

Figure 6.12 – Calculating the epoch value

As shown in the preceding screenshot, this epoch value is `1566880492`.

5. Now, go back to `https://jwt.io` and change the payload as shown in the following code block:

```
{
    "role": "atm_user",
    "exp": 1566880492
}
```

After the preceding lines are executed, we will get a new token that will expire after 5 minutes as shown in *Figure 6.13*:

Figure 6.13 – Setting the expiration for tokens

6. As shown in the following code block, when we access the new token after the 5 minutes, an expiration message will show and PostgREST returns an HTTP 401 Unauthorized error:

```
[centos@ip-172-31-95-213 ~]$ export
TOKEN="eyJhbGciOiJIUzI1NiIsInR5cCI6IkpXVCJ9.eyJyb2xlIjoiYXRtX3VzZXI
iLCJleHAiOjE1NjY4ODA0OTJ9.Imw7-
Mbxejo9SCLOFGfAG01R166fZ6ujhpEIlqvvlig"
[centos@ip-172-31-95-213 ~]$ curl
http://localhost:3000/ATM%20locations -X POST -H "Authorization:
Bearer $TOKEN" -H "Content-Type: application/json" -d
'{"BankName":"Test Bank 2","Address":"100 Test way","County":"New
York","City":"New York","State":"NY","ZipCode":10272}'
{"message":"JWT expired"}
[centos@ip-172-31-95-213 ~]$
```

7. In addition to the expiration tokens, now we will demonstrate how to block a specific user by their email, for example, not.good@mypostgrest.com. We will make the payload JSON as shown:

```
{
   "role": "atm_user",
   "email": "not.good@mypostgrest.com"
}
```

Once the preceding lines are executed, we obtain the following token:

```
eyJhbGciOiJIUzI1NiIsInR5cCI6IkpXVCJ9.eyJyb2xlIjoiYXRtX3VzZXIiLCJlbW
FpbCI6Im5vdC5nb29kQG15cG9zdGdyZXN0LmNvbSJ9.Alz9Wm7oQ4igcZA9mr-
OjgwPJ_d3PisvmKAnb29xLMQ
```

Therefore, the preceding token is considered a bad token for an ATM user with the forbidden not.good@mypostgrest.com email. Whenever there are any new requests using the preceding token, they should be the requests from the bad user that we are trying to block.

8. Next, we will create the following function by pgAdmin. This function will receive the authorized token of a new request, then find out the user email linked to that token. If the extracted email was the bad one, then we will raise an exception to block the user request:

```
create schema auth;
grant usage on schema auth to web_anon, atm_user;

create or replace function auth.check_token() returns void
    language plpgsql
    as $$
```

```
begin
    if current_setting('request.jwt.claim.email', true) =
'not.good@mypostgrest.com' then
        raise insufficient_privilege using hint = 'Nope, we are on to
you';
 end if;
end $$;
```

9. Next, we will stop the PostgREST server by pressing *Ctrl + C* on the console, and then update the `tutorial.conf` file and specify the new function as shown:

```
pre-request = "auth.check_token"
```

An actual view of the preceding command on the console is shown in *Figure 6.14*:

Figure 6.14 – Setting the revocation for tokens

10. Then we restart PostgREST and call the `tutorial.conf` file as shown:

```
[root@ip-172-31-95-213 postgrest]# postgrest tutorial.conf
```

11. Then, as shown, if now we use the blocked token from the specific user to insert, we will get a 403 Forbidden error because that user is not allowed:

```
[centos@ip-172-31-95-213 ~]$ export
BADTOKEN="eyJhbGciOiJIUzI1NiIsInR5cCI6IkpXVCJ9.eyJyb2xlIjoiYXRtX3Vz
ZXIiLCJlbWFpbCI6Im5vdC5nb29kQG15cG9zdGdyZXN0LmNvbSJ9.Alz9Wm7oQ4igcZ
A9mr-OjgwPJ_d3PisvmKAnb29xLMQ"
```

```
[centos@ip-172-31-95-213 ~]$ curl
http://localhost:3000/ATM%20locations -X POST -H "Authorization:
Bearer $BADTOKEN" -H "Content-Type: application/json" -d
'{"BankName":"Test Bank 2","Address":"100 Test way","County":"New
York","City":"New York","State":"NY","ZipCode":10272}'
{"hint":"Nope, we are on to
you","details":null,"code":"42501","message":"insufficient_privileg
e"}
[centos@ip-172-31-95-213 ~]$
```

The previous `pre-request = "auth.check_token"` setting has invoked the
`check_token` function to execute this function before PostgREST can accept the insertion.
Because the token that is sent to PostgREST in this case includes the
`not.good@mypostgrest.com` email, the `insufficient_privilege` exception is then
raised to block this POST request.

PostgREST administration

We will now set up PostgREST so that whenever we start our ec2 instance, the API service
will automatically start:

1. We will configure PostgREST as a service with the help of the following
   command:

   ```
   [root@ip-172-31-95-213 postgrest]# mkdir /etc/postgrest
   [root@ip-172-31-95-213 postgrest]# vi /etc/postgrest/config
   ------------------------------------------
    db-uri = "postgres://dba:bookdemo@atm.ck5074bwbilj.us-
   east-1.rds.amazonaws.com/atm"
    db-schema = "public"
    db-anon-role = "web_anon"
    db-pool = 10
    server-host = "127.0.0.1"
    server-port = 3000
    jwt-secret = "DFZ49GQGubpzcSbt3t2uMIiBF6pU4PJ8"
   ------------------------------------------
   ```

   After the preceding command is executed, the console will look as shown in
   Figure 6.15:

```
root@ip-172-31-95-213:/usr/local/src/postgrest                          —    □    ×
[root@ip-172-31-95-213 postgrest]#
[root@ip-172-31-95-213 postgrest]# vi /etc/postgrest/config
db-uri = "postgres://dba:bookdemo@atm.ck5074bwbilj.us-east-1.rds.amazonaws.com/atm"
db-schema = "public"
db-anon-role = "web_anon"
db-pool = 10

server-host = "127.0.0.1"
server-port = 3000

jwt-secret = "DFZ49GQGubpzcSbt3t2uMIiBF6pU4PJ8"
```

Figure 6.15 – Configuring PostgREST as a service

2. Then create the systemd service file by using the following command:

```
[root@ip-172-31-95-213 postgrest]# ln -s /usr/local/bin/postgrest
/bin/postgrest
[root@ip-172-31-95-213 postgrest]# vi
/etc/systemd/system/postgrest.service
------------------------------------------
 [Unit]
 Description=REST API for any Postgres database
 After=postgresql.service
 [Service]
 ExecStart=/bin/postgrest /etc/postgrest/config
 ExecReload=/bin/kill -SIGUSR1 $MAINPID
 [Install]
 WantedBy=multi-user.target
------------------------------------------
```

After the preceding command is executed, the console will look as shown in *Figure 6.16*:

Figure 6.16 – Creating the systemd service file

3. Now we can enable the `postgrest` service at boot time and start it with the following:

```
[root@ip-172-31-95-213 postgrest]# systemctl enable postgrest
 Created symlink from /etc/systemd/system/multi-
user.target.wants/postgrest.service to
/etc/systemd/system/postgrest.service.
[root@ip-172-31-95-213 postgrest]# systemctl start postgrest
```

From now on, whenever the CentOS `ec2` instance starts, then the API is also running on port `3000`.

PostgREST on TimescaleDB

TimescaleDB is an open source database for time series data; we first heard about TimescaleDB when we investigated standalone PostgREST because it had TimescaleDB as a built-in extension. Obviously, PostgREST is also able to create a timing API for TimescaleDB.

Unfortunately, the PostgreSQL RDS does not support TimescaleDB. If we would like to install TimescaleDB for AWS, we will have to use the recently invented Timescale Cloud to connect to AWS. Timescale Cloud can support TimescaleDB for AWS:

1. Please visit the following link to sign up for a new user account at Timescale Cloud: `https://www.timescale.com/cloud-signup`.

2. Please enter your full name, email address, and password for Timescale Cloud:

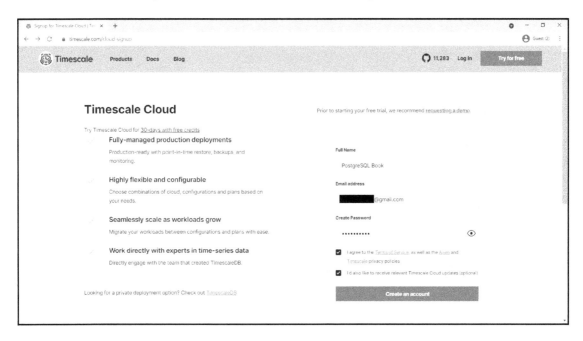

Figure 6.17 – Registering for a new Timescale Cloud user account

3. Please press the **Create an account** button, which will lead you to the **Email confirmed** screen:

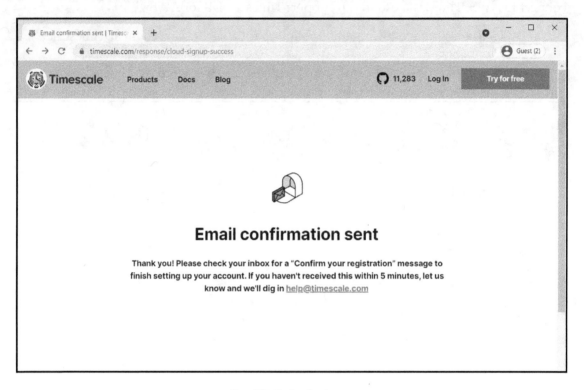

Figure 6.18 – Email confirmed screen

4. Please find the Timescale Cloud registration email from your inbox and open the verification link supplied inside that email. You will see the acceptance of your new Timescale account as shown here:

Figure 6.19 – Acceptance of the new Timescale account

5. After that, you can visit the Timescale Cloud portal and log in with your username and password. You can also use the following link: `http://portal.timescale.cloud/`.

6. Please click on the **+ Create a new service** button:

Figure 6.20 – Timescale Cloud portal

7. On the next screen, please select the TimescaleDB service for PostgreSQL 12 for step 1:

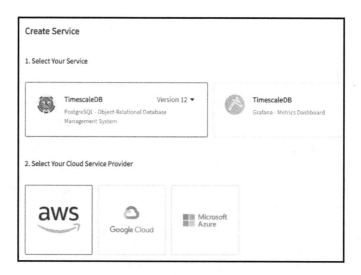

Figure 6.21 – Selecting the TimescaleDB service through AWS

In step 2, please select AWS to connect to TimescaleDB.

8. For step 3, we are selecting the network zone named **timescale-aws-us-east-1** in the United States:

Figure 6.22 – Selecting a network zone for AWS

9. In step 4, we can select a development plan for our deployment:

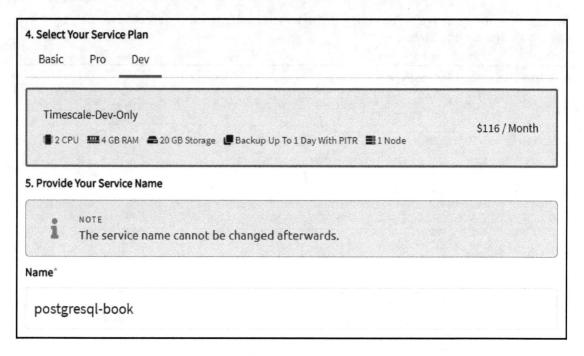

4. Select Your Service Plan

Basic   Pro   Dev

Timescale-Dev-Only

🖥 2 CPU   ▦ 4 GB RAM   💾 20 GB Storage   💾 Backup Up To 1 Day With PITR   ▤ 1 Node

$116 / Month

5. Provide Your Service Name

> ℹ **NOTE**
> The service name cannot be changed afterwards.

Name*

postgresql-book

Figure 6.23 – Selecting a development plan

Then, please enter your service name.

10. Please review the service details for the last time, and then click on the **Create Service** button:

Name

postgresql-book

Service

TimescaleDB 12

Cloud

Amazon Web Services

Region

**United States, Virginia - Timescale /
AWS: N. Virginia**

Plan

timescale-dev-only

- 2 CPU
- 4 GB RAM
- 20 GB Storage
- Backup Up To 1 Day With PITR
- 1 Node

Estimated Monthly Price*

$116

*Estimated monthly price is based on 730 hours of
usage.

Create Service

Figure 6.24 – Revision of your service

11. Timescale Cloud will bring you back to the services screen; you will have to wait a few minutes until your new service is ready:

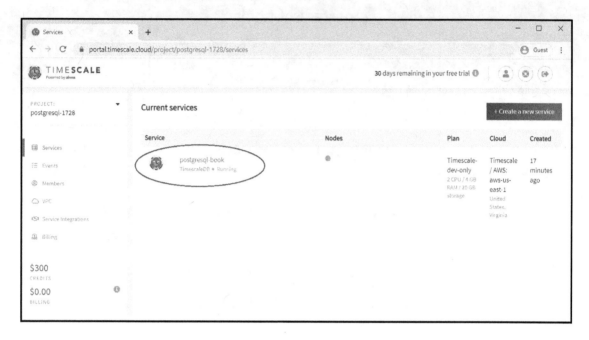

Figure 6.25 – Your new service is now ready

12. Please click on the name of the **postgresql-book** service highlighted in the preceding figure. You will reach the overview page of the service details, as shown here:

Figure 6.26 – Service overview

Please copy the following TimescaleDB details:

- **Database Name**: `defaultdb`
- **Host**: `postgresql-book-postgresql-1728.a.timescaledb.io`
- **Port**: `20153`
- **User**: `tsdbadmin`

13. Please launch pgAdmin and set up a new connection to your new TimescaleDB service. We right-click on **Servers** and then select **Create** | **Server...**:

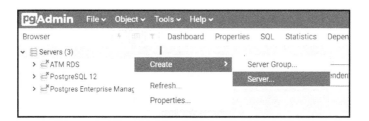

Figure 6.27 – Connecting TimescaleDB by pgAdmin

14. Please enter a new name for your pgAdmin server, and then select the **Connection** tab:

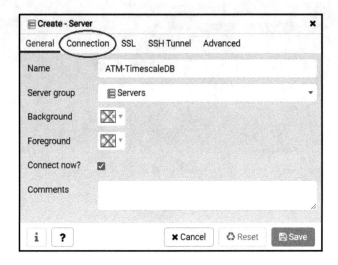

Figure 6.28 – Connecting TimescaleDB by pgAdmin (cont.)

15. Please fill in your values copied from Timescale Cloud here:

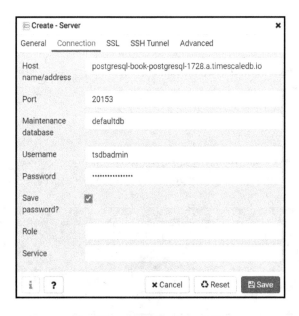

Figure 6.29 – pgAdmin TimescaleDB parameters

We have the following parameters here:

- **Host name/address**: `postgresql-book-postgresql-1728.a.timescaledb.io`
- **Port**: `20153`
- **Maintenance database**: `defaultdb`
- **Username**: `tsdbadmin`
- **Password**: Copied from Timescale Cloud

16. Please click the **Save** button, then we can expand the new `ATM-TimescaleDB` server as follows:

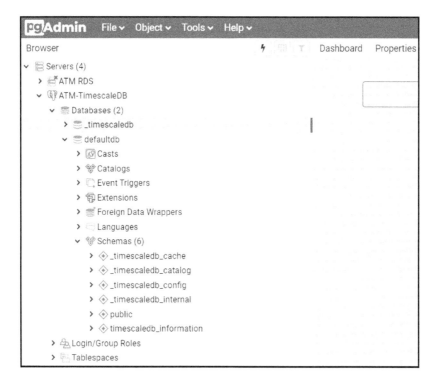

Figure 6.30 – ATM-TimescaleDB

17. Because TimescaleDB is a PostgreSQL extension for timing data – the kind of data that will flexibly change over time – we can choose a banking topic of exchange rates. The exchange rates are changing every hour and this kind of data from banking transactions can be used very well to illustrate timing in TimescaleDB.

Therefore, please open the following GitHub link to create new tables for banking exchange rates of currencies: `https://github.com/lequanha/PostgreSQL-12-Development-and-Administration-Projects/blob/master/Chapter 6/FX-Schema.sql`.

18. Please copy the SQL script from the preceding GitHub link as in the following screenshot:

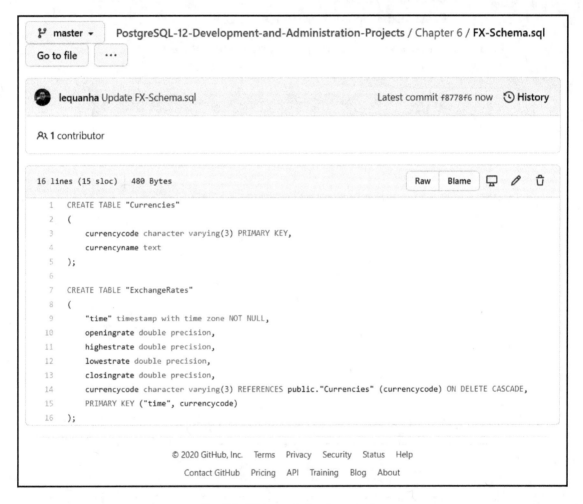

```
1   CREATE TABLE "Currencies"
2   (
3       currencycode character varying(3) PRIMARY KEY,
4       currencyname text
5   );
6
7   CREATE TABLE "ExchangeRates"
8   (
9       "time" timestamp with time zone NOT NULL,
10      openingrate double precision,
11      highestrate double precision,
12      lowestrate double precision,
13      closingrate double precision,
14      currencycode character varying(3) REFERENCES public."Currencies" (currencycode) ON DELETE CASCADE,
15      PRIMARY KEY ("time", currencycode)
16  );
```

Figure 6.31 – GitHub schema for the exchange rates tables

19. We select the expected location as **ATM-TimescaleDB** | **Databases** | **defaultdb** | **Schemas** | **public,** then please paste the preceding SQL script into the Query Editor of your pgAdmin:

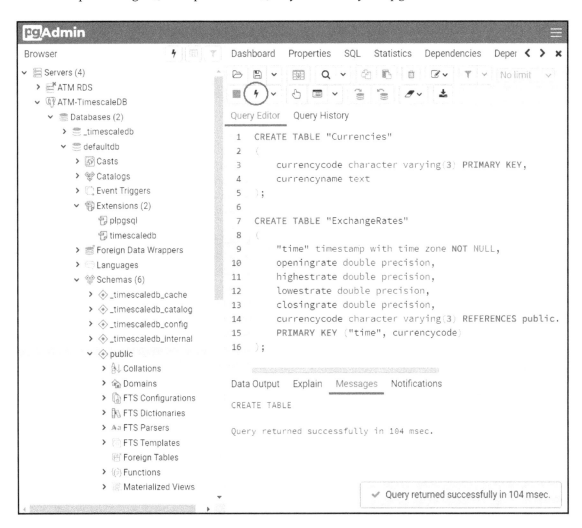

Figure 6.32 – Creating the ExchangeRates and Currencies tables in pgAdmin

Finally, press the ⚡ (Execute/Refresh (*F5*)) icon to create our new tables.

20. At first, we should convert the `ExchangeRates` table into a hypertable with just timing partitioning on the `time` column. Then, we add an additional partition key on the `currencycode` column with three partitions with the following SQL script:

```
SELECT create_hypertable('"ExchangeRates"', 'time');
SELECT add_dimension('"ExchangeRates"', 'currencycode',
number_partitions => 3);
```

Hypertables in TimescaleDB are designed similarly to PostgreSQL tables. Developers prefer to use hypertables to improve the read and write performance of the database. In comparison to a traditional table, a big data hypertable will be able to perform up to 15 times quicker, while a small data hypertable will still insert, update, delete, and select data at the same speed, causing no additional overhead at all.

21. We can copy the preceding script into the Query Editor of pgAdmin:

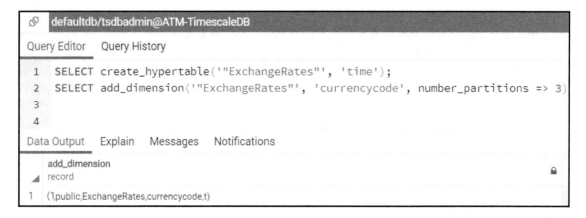

Figure 6.33 – The ExchangeRates hypertable

22. Please open the following GitHub link that contains the SQL data script for our currencies: `https://github.com/lequanha/PostgreSQL-12-Development-and-Administration-Projects/blob/master/Chapter 6/Currencies.sql`.

We can paste the script from GitHub into the Query Editor of pgAdmin to execute as follows:

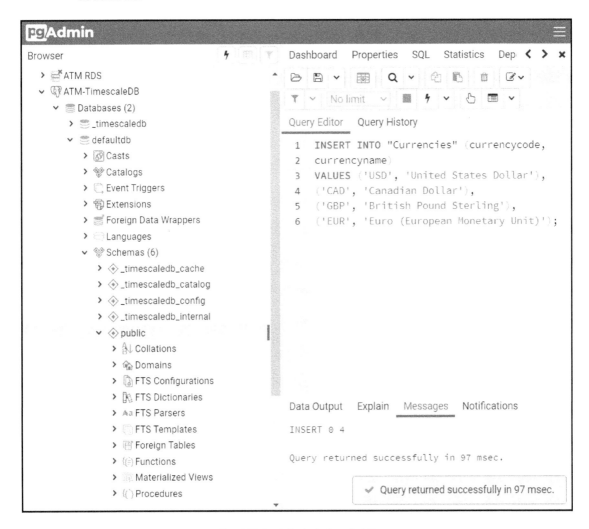

Figure 6.34 – Sample data for the Currencies table

23. We are going to insert exchange rates from 1 USD to CAD, 1 USD to GBP, and 1 USD to EUR. The exchange rates are stored every hour for the first 6 months from January to the end of June 2020.

Please use the following GitHub link: `https://github.com/lequanha/`
`PostgreSQL-12-Development-and-Administration-Projects/blob/master/`
`Chapter 6/USDCAD_Jan2020_Jun2020.sql`.

When you open this link, you will see the INSERT script of 4,368 records of USD-CAD exchange rates:

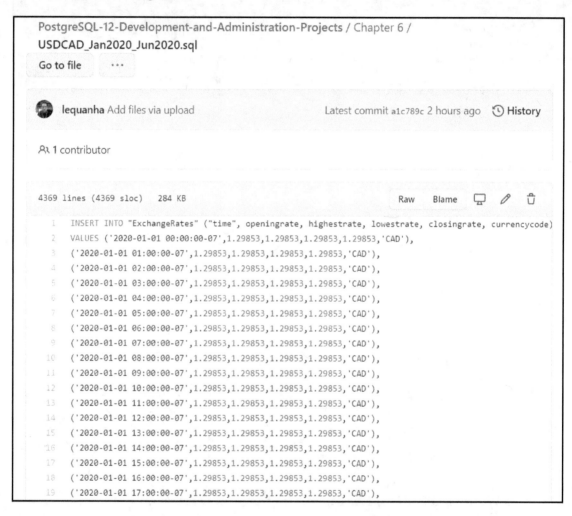

Figure 6.35 – Hourly USD-CAD exchange rates for the first half of 2020

We requested usage permission for all of our data from Dukascopy and they have granted their acceptance to use their data for this book from here: `https://www.dukascopy.com/swiss/english/marketwatch/historical/`.

24. The preceding GitHub INSERT script can be copied into pgAdmin to execute as follows:

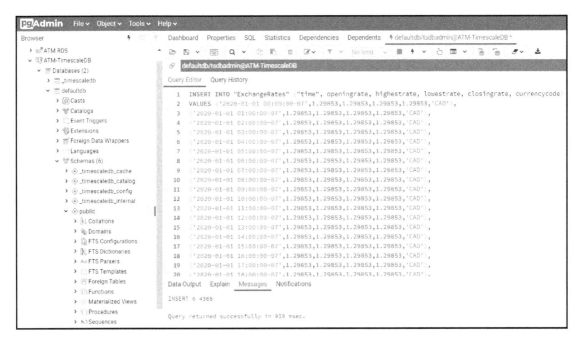

Figure 6.36 – Hourly USD-CAD exchange rates are populated into our TimescaleDB

25. Similarly, we next import USD-GBP exchange rates and USD-EUR rates for the first 6 months of 2020 from the following two GitHub links:

- `https://github.com/lequanha/PostgreSQL-12-Development-and-Administration-Projects/blob/master/Chapter 6/USDGBP_Jan2020_Jun2020.sql`

- `https://github.com/lequanha/PostgreSQL-12-Development-and-Administration-Projects/blob/master/Chapter 6/USDEUR_Jan2020_Jun2020.sql`

26. Please select the `ExchangeRates` table on the left panel, and then 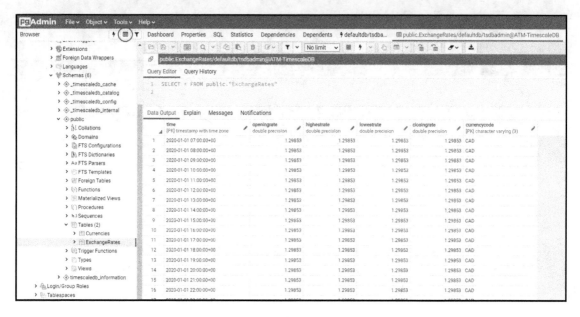 ss the (View Data) icon on the toolbar to display the final hourly data of all USD-CAD, USD-GBP, and USD-EUR rates inside our TimescaleDB now:

Figure 6.37 – Successful exchange rate import

27. Please use this timing SELECT query to filter weekly exchange rates from our hourly table as follows:

```
SELECT time_bucket('1 week', time) as period,
       last(closingrate, time) AS last_closing_rate,
       currencycode
FROM "ExchangeRates"
GROUP BY period, currencycode
ORDER BY period DESC;
```

This is the screenshot from pgAdmin:

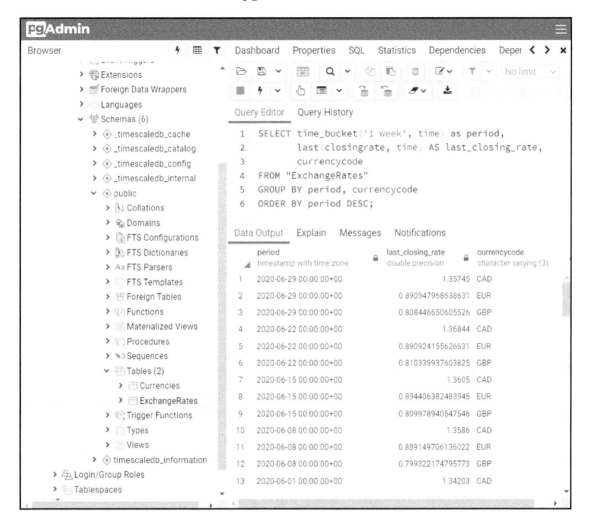

Figure 6.38 – Weekly exchange rates

Hence, for the latest week on Monday, June 29th, 1 USD was equal to 1.357 CAD, 0.891 EUR, or 0.808 GBP. The previous week before that, on Monday, June 22nd, 1 USD was equal to 1.368 CAD, 0.891 EUR, or 0.810 GBP, and more.

28. If we would like to find out the first day of each exchange rate inside our TimescaleDB, we can use the `min` function of time data as follows:

```
SELECT ci.currencycode, min(c.time)
FROM "Currencies" ci JOIN "ExchangeRates" c ON ci.currencycode =
c.currencycode
GROUP BY ci.currencycode
ORDER BY min(c.time) DESC;
```

Please copy the preceding SELECT query into pgAdmin:

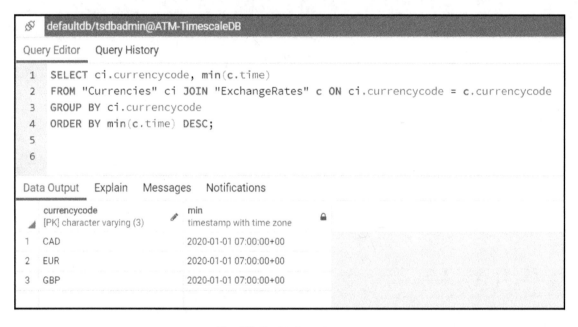

Figure 6.39 – Start date of our exchange rates

Hence, all of the three types of USD-CAD, USD-EUR, and USD-GBP exchange rates start in our TimescaleDB from January 1st, 2020.

29. Now please practice the creation of a new view to filter daily exchange rates from our original hourly data, as follows:

```
CREATE VIEW ExchangeRates_Daily
WITH (timescaledb.continuous)
AS
SELECT
    time_bucket('1 day', time) AS period,
    first(openingrate, time) AS openingrate,
    max(highestrate) AS highestrate,
```

```
      min(lowestrate) AS lowestrate,
      last(closingrate, time) AS closingrate,
      currencycode
FROM "ExchangeRates"
GROUP BY period, currencycode;
```

The data frequency is 1 day for the daily filter. There are the **first** timing functions to get the **earliest** exchange rate within one day, while the **last** function will select the latest exchange rate of that day:

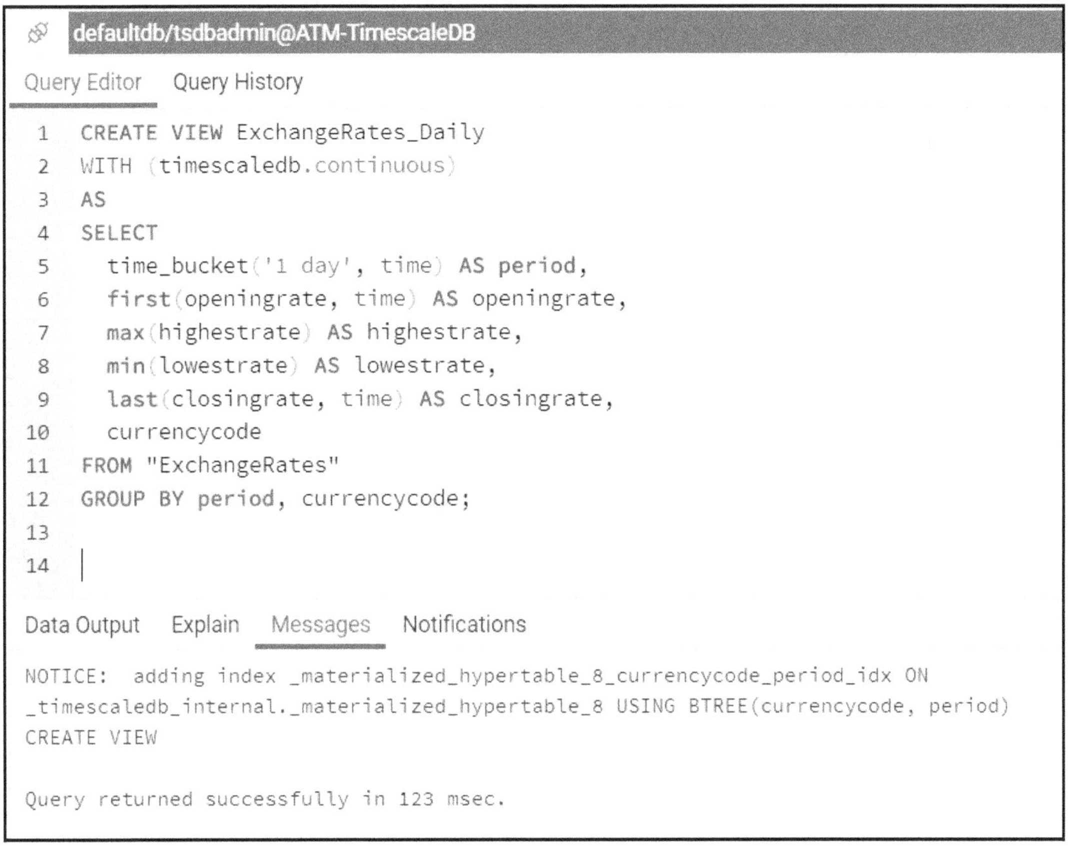

```
     defaultdb/tsdbadmin@ATM-TimescaleDB

Query Editor    Query History

 1    CREATE VIEW ExchangeRates_Daily
 2    WITH (timescaledb.continuous)
 3    AS
 4    SELECT
 5      time_bucket('1 day', time) AS period,
 6      first(openingrate, time) AS openingrate,
 7      max(highestrate) AS highestrate,
 8      min(lowestrate) AS lowestrate,
 9      last(closingrate, time) AS closingrate,
10      currencycode
11    FROM "ExchangeRates"
12    GROUP BY period, currencycode;
13
14    |

Data Output   Explain   Messages   Notifications

NOTICE:  adding index _materialized_hypertable_8_currencycode_period_idx ON
_timescaledb_internal._materialized_hypertable_8 USING BTREE(currencycode, period)
CREATE VIEW

Query returned successfully in 123 msec.
```

Figure 6.40 – Creation of the ExchangeRates_Daily view

30. Please select the view on the left panel by going to **Schemas** | **public** | **Views** | **exchangerates_daily**, and then press the ⊞ (View Data) icon on the toolbar to display the daily data of exchange rates:

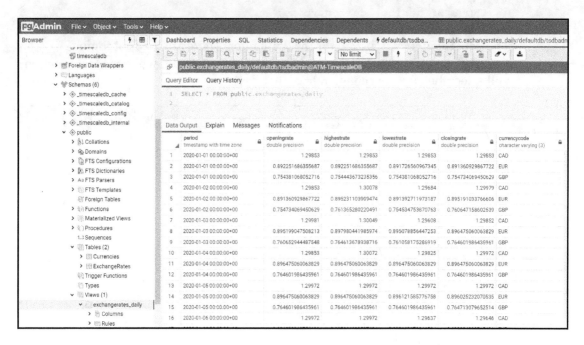

Figure 6.41 – Results of the ExchangeRates_Daily view

Hence, on New Year's Day, January 1st, at the opening of the day, 1 USD could exchange for 1.299 CAD, 0.892 EUR, or 0.754 GBP; at the closing of the same day, 1 USD could exchange for 1.299 CAD, 0.891 EUR, or 0.755 GBP. Similarly, for January 2nd at closing time, 1 USD converted to less than 1.300 CAD, 0.895 EUR, or 0.761 GBP, and so on.

31. In order to check the status of the view materialization job, please use this TimescaleDB query:

```
SELECT * FROM timescaledb_information.continuous_aggregate_stats;
```

The result is shown:

Figure 6.42 – Materialization job of the ExchangeRates_Daily view

It means that the last time the `ExchangeRates_Daily` view was materialized was at 01:02 a.m. on July 22nd, completing after 9 milliseconds (which is very fast!).

Summary

In this chapter, we have implemented PostgREST step by step to create an automatic API service for PostgreSQL version 12. By working through this chapter, developers have also practiced Docker usage, Docker and SQL, PostgREST first execution, database users and roles, and JSON Web Token. We also learned how to run PostgREST as a service.

The second part of the chapter focused on introducing PostgreSQL timing data with TimescaleDB from Timescale Cloud, a new extension related to PostgREST. This chapter concludes the PostgreSQL development part of the book.

In the next chapters, we will focus on PostgreSQL administration.

Section 3 - Administration in PostgreSQL

This section handles the administration of PostgreSQL using DevOps, including setting up high-availability PostgreSQL clusters, setting up New Relic to monitor a PostgreSQL database, carrying out performance tests on a PostgreSQL database with PGBench and JMeter, and using PostgreSQL testing frameworks.

This section contains the following chapters:

- Chapter 7, *PostgreSQL with DevOps for Continuous Delivery*
- Chapter 8, *PostgreSQL High Availability Clusters*
- Chapter 9, *High-Performance Team Dashboards Using PostgreSQL and New Relic*
- Chapter 10, *Testing the Performance of Our Banking App with PGBench and JMeter*
- Chapter 11, *Test Frameworks for PostgreSQL*
- Chapter 12, *APPENDIX - PostgreSQL among the Other Current Clouds*

PostgreSQL with DevOps for Continuous Delivery

In this chapter, you will learn how to set up DevOps tasks for PostgreSQL databases. DevOps is no longer just about breaking the gap between developers and operations. It is, rather, a software development methodology based on the integration between developers and system administrators. Database changes are always a headache because whether they are improvements, fixes, or new features, these will alter the current environment and perhaps create a new one that could be heaven or a nightmare and therefore deserves to be taken seriously in your DevOps implementation. Hence, we will explain a different perspective for automating database changes. We will not only see why database changes can be difficult, but will also go through some real-world examples of how DevOps can simplify the process.

In the project in this chapter, we will use DevOps tools to implement the PostgreSQL 12 database of our banking ATM machine locations by using virtual machines and popular CI/CD tools.

The following topics will be covered in the chapter:

- Setting up PostgreSQL using Vagrant and VirtualBox

- Creating the Puppet module for PostgreSQL

- Working with Jenkins and PostgreSQL

- Creating an Ansible playbook for PostgreSQL

- Managing PostgreSQL by Terraform

Technical requirements

The examples given in this chapter will take developers around 20-24 hours'work to implement DevOps for the ATM PostgreSQL database.

The code files for this chapter are available at the following link: `https://github.com/PacktPublishing/Developing-Modern-Database-Applications-with-PostgreSQL/tree/master/Chapter07`.

Setting up PostgreSQL using Vagrant and VirtualBox

In this chapter, we will use Vagrant and VirtualBox. Both tools are widely used and accepted for DevOps.

The idea behind these useful tools is to quickly create and share environments for developers and testers. We will see how to achieve this in the following paragraphs.

Installing VirtualBox

VirtualBox is a well-known virtualization product for enterprise as well as home use. It can create a VM of another OS such as Windows, Ubuntu, Debian, RHEL, Solaris, and OS/2. In the following steps, we will learn how to use it.

Note that we are using a Windows 10 OS and there we will perform these tasks:

1. The first step is to download the VirtualBox installer from their website: `https://www.virtualbox.org/wiki/Downloads`. Once there, we will download the latest version available at the time of writing this book: `https://download.virtualbox.org/virtualbox/6.1.4/VirtualBox-6.1.4-136177-Win.exe`.

2. Now, we can start to run this VirtualBox installer as an administrator feature of our Windows 10 interface. The installer's Welcome screen will be displayed.

3. After clicking the **Next** button, you'll see a **Custom Setup** window:

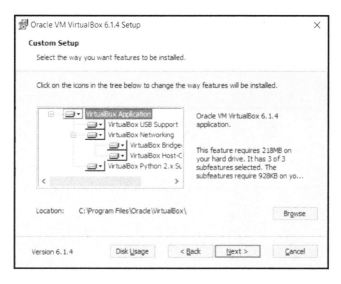

Figure 7.1 – VirtualBox features

4. Now, select some options related to features in the Windows interface and how to access VirtualBox from it:

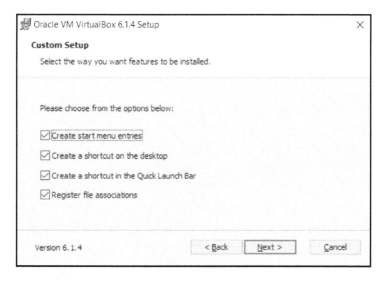

Figure 7.2 – Custom Setup

5. The Network reset warning option will be displayed in the next window. Select **Yes** to proceed with the installation:

Figure 7.3 – Network Interfaces

6. Before the last step, we should confirm **Yes** to proceed with the installation:

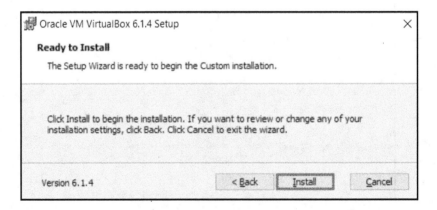

Figure 7.4 – Ready to Install VirtualBox

7. After waiting a short time, the installation will be complete and you can click on the **Finish** button to finalize:

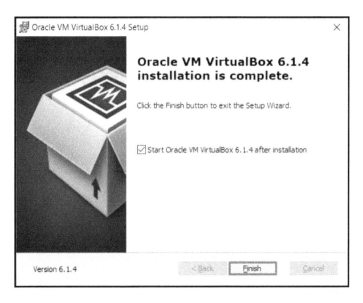

Figure 7.5 – Installation complete

8. Finally, VirtualBox Manager will be shown as follows:

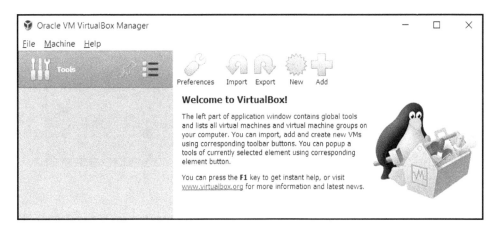

Figure 7.6 – VirtualBox is starting

Now that VirtualBox has been installed, we can proceed to install Vagrant so that any VMs generated from the Vagrant command-line interface can be managed with our VirtualBox tool.

Installing Vagrant

Vagrant is an open source tool for deploying and managing software development environments. It isolates the dependencies and configuration from your projects into environments called **virtual machines** (**VMs**). You could easily reproduce similar VMs on other computers by using the same configuration as on your computer:

1. The first step is to download Vagrant from their website: `https://www.vagrantup.com/downloads.html`.

 The current version at the time of writing this book is `vagrant_2.2.7`: `https://releases.hashicorp.com/vagrant/2.2.7/vagrant_2.2.7_x86_64.msi`.

2. We will now execute the Vagrant installer by using the **Run as administrator** feature and clicking the **Run anyway** button:

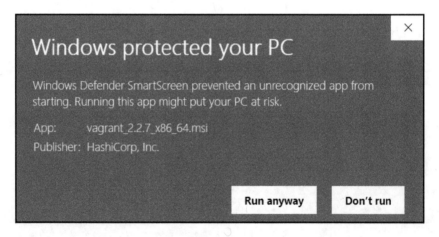

Figure 7.7 – Starting the installation of Vagrant

3. The setup wizard is presented at this point and we then click on the **Next** button:

Figure 7.8 – Vagrant Setup Wizard

4. The end user license agreement is shown and we must accept it in order to continue:

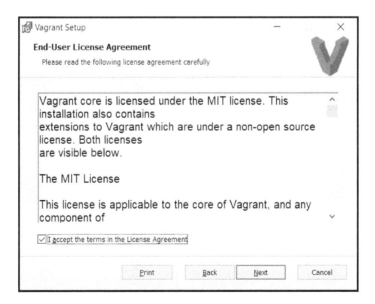

Figure 7.9 – Vagrant License Agreement

5. We can either choose the destination folder or we just select the default option:

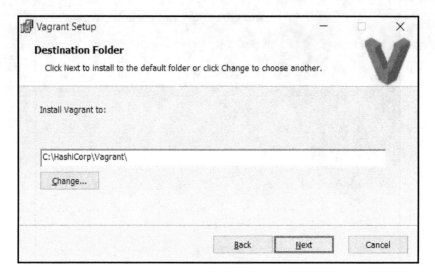

Figure 7.10 – Selecting the destination folder

6. Now we're ready to install Vagrant. Click on the **Install** button:

Figure 7.11 – Ready to install Vagrant

7. A progress bar will show how the installation process is proceeding:

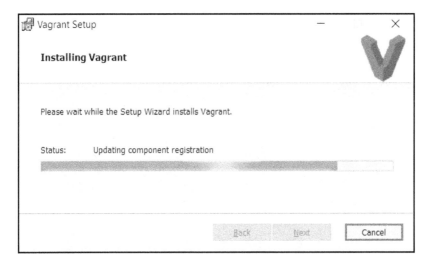

Figure 7.12 – Installing Vagrant

8. In the end, we will see that the installation has concluded successfully:

Figure 7.13 – Completed Vagrant installation

9. A request to restart your computer will pop up. Click on the **Yes** button:

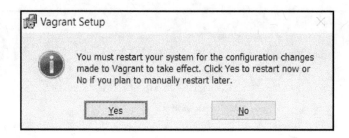

Figure 7.14 – Restarting Vagrant

Once your computer restarts, Vagrant is ready to create VMs. Each VM is configured with a Vagrant box to select which OS you would like to host on the VM.

Selecting a Vagrant box

Now that both VirtualBox and Vagrant are installed on your physical computer, it is time to browse the list of available boxes for Vagrant. Open the list of base boxes from this link: `https://app.vagrantup.com/boxes/search`:

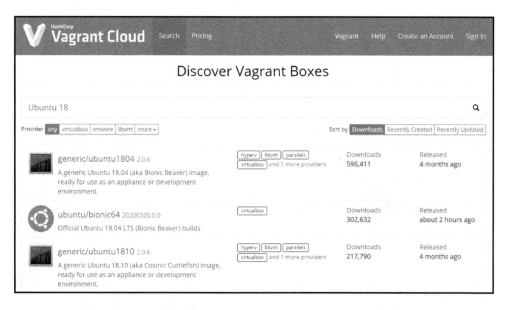

Figure 7.15 – List of all available Vagrant boxes

Note that a base box is a template of a VM and is defined on the Vagrant site as follows:"Boxes are the package format for Vagrant environments. A box can be used by anyone on any platform that Vagrant supports to bring up an identical working environment."

We can write "Ubuntu 18" to list all of the Ubuntu 18 boxes because we next use the bento/ubuntu-18.04 box to create a few Ubuntu 18 VMs.

Spinning up Ubuntu server 18.04 with Vagrant

Because of the new releases of VirtualBox 6.1, unfortunately, Vagrant 2.2.6 has not yet updated on time to fully work with this new release. I will show here how to fix the issue and then Windows 10 will able to work for the two latest versions of Vagrant and VirtualBox:

1. Add this line to the file on your Windows 10 machine: `C:\HashiCorp\Vagrant\embedded\gems\2.2.7\gems\vagrant-2.2.7\plugins\providers\virtualbox\plugin.rb`:

```
module Driver
    autoload :Meta, File.expand_path("../driver/meta", __FILE__)
    autoload :Version_4_0,
File.expand_path("../driver/version_4_0", __FILE__)
    autoload :Version_4_1,
File.expand_path("../driver/version_4_1", __FILE__)
    autoload :Version_4_2,
File.expand_path("../driver/version_4_2", __FILE__)
    autoload :Version_4_3,
File.expand_path("../driver/version_4_3", __FILE__)
    autoload :Version_5_0,
File.expand_path("../driver/version_5_0", __FILE__)
    autoload :Version_5_1,
File.expand_path("../driver/version_5_1", __FILE__)
    autoload :Version_5_2,
File.expand_path("../driver/version_5_2", __FILE__)
    autoload :Version_6_0,
File.expand_path("../driver/version_6_0", __FILE__)
    autoload :Version_6_1,
File.expand_path("../driver/version_6_1", __FILE__)
  end
```

2. Then, add this line to the file: `C:\HashiCorp\Vagrant\embedded\gems\2.2.7\gems\vagrant-2.2.7\plugins\providers\virtualbox\driver\meta.rb`:

```
driver_map = {
    "4.0" => Version_4_0,
    "4.1" => Version_4_1,
    "4.2" => Version_4_2,
    "4.3" => Version_4_3,
    "5.0" => Version_5_0,
    "5.1" => Version_5_1,
    "5.2" => Version_5_2,
    "6.0" => Version_6_0,
    "6.1" => Version_6_1,
}
```

3. Finally, create a new file called `version_6_1.rb` in the
   `C:\HashiCorp\Vagrant\embedded\gems\2.2.6\gems\vagrant-2.2.6\plu`
   `gins\providers\virtualbox\driver\` directory:

```
require File.expand_path("../version_6_0", __FILE__)
    module VagrantPlugins
        module ProviderVirtualBox
            module Driver
                # Driver for VirtualBox 6.1.x
                class Version_6_1 < Version_6_0
                    def initialize(uuid)
                        super
                        @logger =
Log4r::Logger.new("vagrant::provider::virtualbox_6_1")
                    end
                end
            end
        end
    end
```

4. Now, Vagrant 2.2.6 will work alright for VirtualBox 6.1. Proceed with the
   following statements in Windows PowerShell to start up Ubuntu 18.04 for the
   PostgreSQL-Master VM. Open PowerShell as an administrator:

```
PS C:\Windows\system32> mkdir C:\Projects
PS C:\Windows\system32> cd \Projects
PS C:\Projects> mkdir Vagrant
PS C:\Projects> cd Vagrant
PS C:\Projects\Vagrant> mkdir PostgreSQL-Master
PS C:\Projects\Vagrant> cd PostgreSQL-Master
```

5. If you have not yet opened VirtualBox, remember to launch it now.

6. Now, we start up the PostgreSQL VM:

```
PS C:\Projects\Vagrant\PostgreSQL-Master> bcdedit /set
hypervisorlaunchtype off
PS C:\Projects\Vagrant\PostgreSQL-Master> vagrant init
bento/ubuntu-18.04
PS C:\Projects\Vagrant\PostgreSQL-Master> vagrant up --provider
virtualbox
```

If you are successful in launching a new Ubuntu VM, you will observe a similar screenshot to the following:

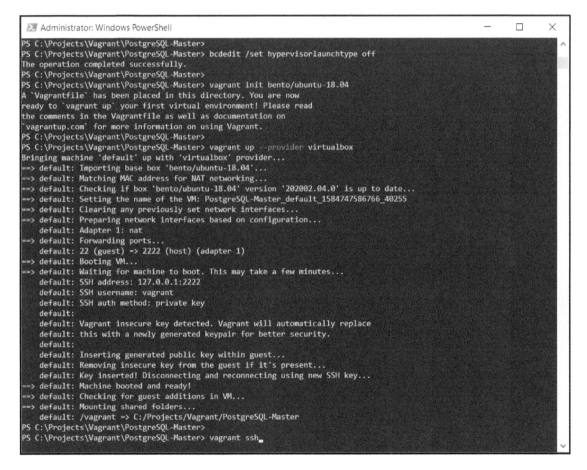

Figure 7.16 – Using Vagrant to create our new VM

7. Once the Vagrant setup statement has been completed, you will see that the new VM is automatically added to VirtualBox:

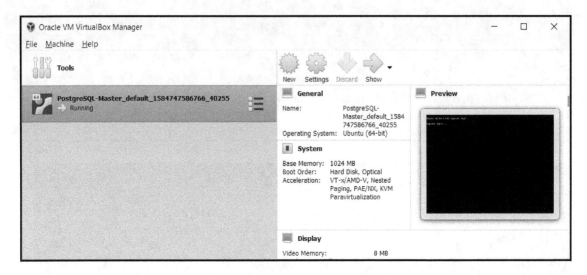

Figure 7.17 – Our new Ubuntu 18 instance on VirtualBox

8. Now we can `ssh` into our new Vagrant VM:

```
PS C:\Projects\Vagrant\PostgreSQL-Master> vagrant ssh
```

The preceding Vagrant `ssh` statement does not require user authentication. It will lead you straight to the VM session:

```
==> default: Mounting shared folders...
    default: /vagrant => C:/Projects/Vagrant/PostgreSQL-Master
PS C:\Projects\Vagrant\PostgreSQL-Master>
PS C:\Projects\Vagrant\PostgreSQL-Master> vagrant ssh
Welcome to Ubuntu 18.04.4 LTS (GNU/Linux 4.15.0-76-generic x86_64)

 * Documentation:  https://help.ubuntu.com


 * Documentation:  https://help.ubuntu.com
 * Management:      https://landscape.canonical.com
 * Support:         https://ubuntu.com/advantage

  System information as of Fri Mar 20 23:44:04 UTC 2020

  System load:  0.0               Processes:            93
  Usage of /:   2.5% of 61.80GB   Users logged in:      0
  Memory usage: 11%               IP address for eth0:  10.0.2.15
  Swap usage:   0%

 * Latest Kubernetes 1.18 beta is now available for your laptop, NUC, cloud
   instance or Raspberry Pi, with automatic updates to the final GA release.

     sudo snap install microk8s --channel=1.18/beta --classic

 * Multipass 1.1 adds proxy support for developers behind enterprise
   firewalls. Rapid prototyping for cloud operations just got easier.

     https://multipass.run/

0 packages can be updated.
0 updates are security updates.



This system is built by the Bento project by Chef Software
More information can be found at https://github.com/chef/bento
vagrant@vagrant: $ sudo su
root@vagrant:/home/vagrant#
```

Figure 7.18 – The Vagrant ssh session

You are now connected as a *Vagrant* user, which has proper *sudo* capabilities. In this step, you can basically launch your owned Ubuntu **Virtual Machine** (**VMS**). With all this, we will use the preceding configurations as a basis for our future work involving PostgreSQL and Puppet.

Creating the Puppet module for PostgreSQL

Ubuntu includes PostgreSQL on its repositories by default. To install PostgreSQL on Ubuntu, we can use the `apt-get` (or other `apt-driving`) command, but instead of that, we will modify the Vagrant file to include the shell commands to initialize PostgreSQL through Puppet.

Puppet is a software configuration management tool, but I like to say that it is a powerful infrastructure automation and delivery tool. In the following steps, we will see how to deploy PostgreSQL and how it can be configured easily with the help of this extraordinary tool:

1. Log out of the current vagrant `ssh` session and then destroy your Vagrant environment:

   ```
   vagrant destroy
   ```

   Answer y (=yes) to any questions if prompted:

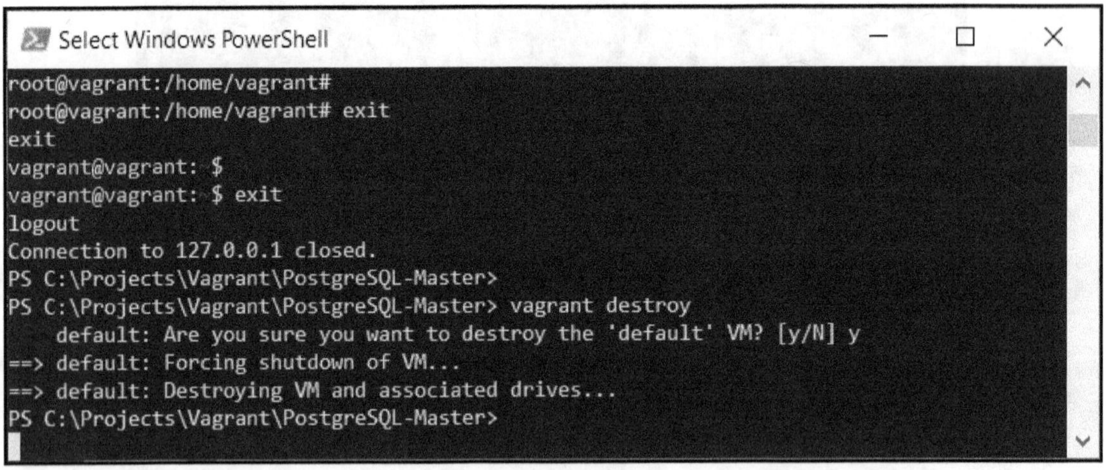

Figure 7.19 – Destroying a VM

2. Change the vagrant file, `C:\Projects\Vagrant\PostgreSQL-Master\Vagrantfile`, and change the commented-out lines (near the bottom of the file) that read as follows:

   ```
   # config.vm.provision "shell", inline: <<-SHELL
   #    apt-get update
   #    apt-get install -y apache2
   # SHELL
   ```

3. Now, replace the previous content with the following text:

```
config.vm.provision "shell", inline: <<-SHELL
    apt-get -y install puppet
    puppet module install puppetlabs-postgresql
SHELL
config.vm.provision "puppet" do |puppet|
    puppet.manifests_path = "puppet/manifests/"
    puppet.manifest_file = "postgresql.pp"
end
```

This VM will install Puppet and the `puppetlabs-postgresql` module. This module offers this VM the ability to manage PostgreSQL databases.

4. Now, set up an IP address for the PostgreSQL VM as shown here. This step is optional:

```
config.vm.network "public_network", ip: "192.168.0.191"
```

5. The entire Vagrant file, `C:\Projects\Vagrant\PostgreSQL-Master\Vagrantfile`, will look like the following code:

```
Vagrant.configure("2") do |config|
    config.vm.box = "bento/ubuntu-18.04"
    config.vm.network "public_network", ip: "192.168.0.191"
    config.vm.provision "shell", inline: <<-SHELL
        apt-get -y install puppet
        puppet module install puppetlabs-postgresql
    SHELL
    config.vm.provision "puppet" do |puppet|
        puppet.manifests_path = "puppet/manifests/"
        puppet.manifest_file = "postgresql.pp"
    end
end
```

6. Save the changes and create the `C:\Projects\Vagrant\PostgreSQL-Master\puppet\manifests` directory as shown here:

```
PS C:\Projects\Vagrant\PostgreSQL-Master> mkdir puppet
PS C:\Projects\Vagrant\PostgreSQL-Master> mkdir puppet\manifests
```

7. Create the `postgresql.pp` text file inside the
   `C:\Projects\Vagrant\PostgreSQL-Master\puppet\manifests` folder with
   the following content:

```
class pgbook {
    class { 'postgresql::globals':
        version => '12',
        manage_package_repo => true,
        encoding => 'UTF8',
    } ->
    class { 'postgresql::server':
        package_ensure => latest,
        ip_mask_allow_all_users => '0.0.0.0/0',
        listen_addresses => '*',
    } ->
    postgresql::server::db { 'atm':
        user => 'dba',
        password => postgresql_password('dba', 'bookdemo'),
        encoding => 'UTF8',
    }
    class { 'postgresql::server::contrib':
        package_ensure => latest,
    }
}
class { pgbook: }
```

This Puppet file is quite readable. It declares a PostgreSQL service with version 12
for the VM. The PostgreSQL service allows connections from any IPs. There is a
database named `atm` and the database username will be `dba` with the password
`bookdemo`. The `postgresql-contrib` extension will be at the latest available
version.

8. At this point, execute `vagrant up` and this will create and configure our
   machines according to our Vagrant file:

```
vagrant up --provider virtualbox
vagrant ssh
```

Once the VM is loaded up, Puppet will be installed through Vagrant and
PostgreSQL 12 will be installed as well:

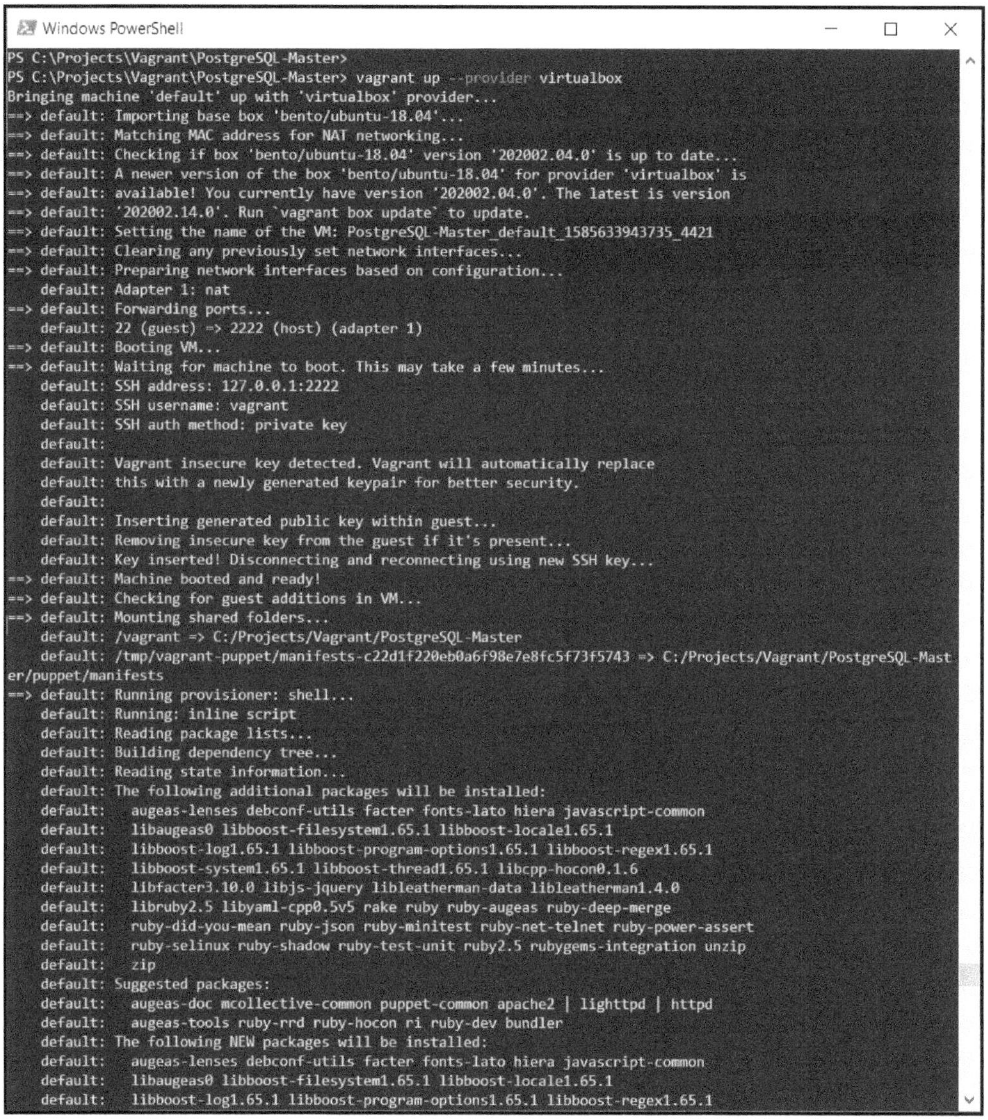

Figure 7.20 – Using Puppet with Vagrant to deploy PostgreSQL 12

9. Run `psql`. If Puppet is correct, you should be able to access PostgreSQL without any issues:

```
vagrant@vagrant:~$ sudo su postgres
postgres@vagrant:/home/vagrant$ psql
psql (12.6 (Ubuntu 12.6-1.pgdg18.04+1))
```

```
Type "help" for help.
postgres=# \x
Expanded display is on.
postgres=# \l *atm*
List of databases
-[ RECORD 1 ]-----+---------------------
Name              | atm
Owner             | postgres
Encoding          | UTF8
Collate           | en_US.UTF-8
Ctype             | en_US.UTF-8
Access privileges | =T/postgres +
                  | postgres=CTc/postgres+
                  | dba=CTc/postgres
postgres=#
```

You are now able to create a new VM of Ubuntu 18 and PostgreSQL 12 by using Vagrant, VirtualBox, and Puppet together. We are using this VM as our PostgreSQL server for the time being and we will not complete our DevOps project with only one VM. In the next section, we will create a second VM, called the Jenkins server, for continuous integration purposes.

Working with Jenkins and PostgreSQL

Jenkins is a free, open source automation server to automate the phases of software development, such as building, testing, and deploying. Jenkins facilitates continuous integration and continuous delivery. We will now practice using Jenkins for PostgreSQL automation:

1. Use this Vagrant box to spin up a Jenkins server on an Ubuntu 18.04 VM:

   ```
   darkwizard242/devopsubuntu1804
   ```

2. Set up the Vagrant file from the `C:\Projects\Vagrant\Jenkins` folder. Open PowerShell as an administrator:

   ```
   PS C:\Windows\system32> mkdir C:\Projects\Vagrant\Jenkins
   PS C:\Windows\system32> cd C:\Projects\Vagrant\Jenkins
   PS C:\Projects\Vagrant\Jenkins> bcdedit /set hypervisorlaunchtype
   off
   PS C:\Projects\Vagrant\Jenkins> vagrant init
   darkwizard242/devopsubuntu1804
   ```

3. Edit the Vagrant file, `C:\Projects\Vagrant\Jenkins\Vagrantfile`, and add the IP address `192.168.0.200`:

```
Vagrant.configure("2") do |config|
    config.vm.box = "darkwizard242/devopsubuntu1804"
    config.vm.network "public_network", ip: "192.168.0.200"
end
```

At this step, the Vagrant file should be configured similar to the one on GitHub.

4. Now, launch the new Jenkins VM:

```
PS C:\Projects\Vagrant\Jenkins> vagrant up --provider virtualbox
```

The following is a screenshot of the Jenkins server by Vagrant and VirtualBox:

Figure 7.21 – Vagrant up for a new Jenkins server

5. Now, visit the following URL, `http://192.168.0.200:8080/`, and the Jenkins login page will display:

Figure 7.22 – Jenkins Welcome page

6. Sign in with a username of `admin` and with `admin` as the password:

Figure 7.23 – Jenkins main page

7. We will use Jenkins to install a replication tool of PostgreSQL called `repmgr` for the PostgreSQL server, `192.168.0.191`.

   Before creating our project on Jenkins, `ssh` into the Jenkins server from PowerShell to copy the `ssh` key into the PostgreSQL server, `192.168.0.191`, so that the Jenkins scripts can later execute remotely on the PostgreSQL server:

   ```
   PS C:\Windows\system32>
   PS C:\Windows\system32> cd C:\Projects\Vagrant\Jenkins
   PS C:\Projects\Vagrant\Jenkins> vagrant up --provider virtualbox
   PS C:\Projects\Vagrant\Jenkins> vagrant ssh

   *** System restart required ***
   Last login: Sat Apr 4 04:49:59 2020 from 10.0.2.2

   vagrant@devopsubuntu1804:~$
   vagrant@devopsubuntu1804:~$ sudo su
   root@devopsubuntu1804:/home/vagrant# su - jenkins
   jenkins@devopsubuntu1804:~$ ssh-copy-id -i
   "/var/lib/jenkins/.ssh/id_rsa" vagrant@192.168.0.191
   /usr/bin/ssh-copy-id: INFO: Source of key(s) to be installed:
   "/var/lib/jenkins/.ssh/id_rsa.pub"
   /usr/bin/ssh-copy-id: INFO: attempting to log in with the new
   key(s), to filter out any that are already installed
   /usr/bin/ssh-copy-id: INFO: 1 key(s) remain to be installed -- if
   you are prompted now it is to install the new keys
   vagrant@192.168.0.191's password: (please type vagrant)
   Number of key(s) added: 1
   Now try logging into the machine, with: "ssh
   'vagrant@192.168.0.191'"
   and check to make sure that only the key(s) you wanted were added.
   jenkins@devopsubuntu1804:~$
   ```

   The `ssh-copy-id` command will copy the `id_rsa` file remotely from the Jenkins server into the PostgreSQL server, `192.168.0.191`, through the user `vagrant`.

8. Now you can make a script remotely from the Jenkins server to execute on the PostgreSQL VM.

Click on **create new jobs** or **New Item**:

Figure 7.24 – Creating a new Jenkins project

9. Enter `RepMgr Installation` and then click on **Freestyle project** and finally **OK**:

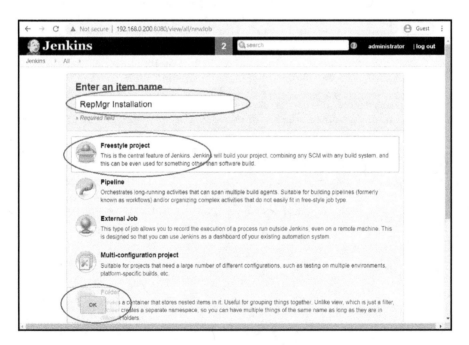

Figure 7.25 – Jenkins' freestyle project

10. Navigate to **Build** and then **Add build step** to select the **Execute shell** option:

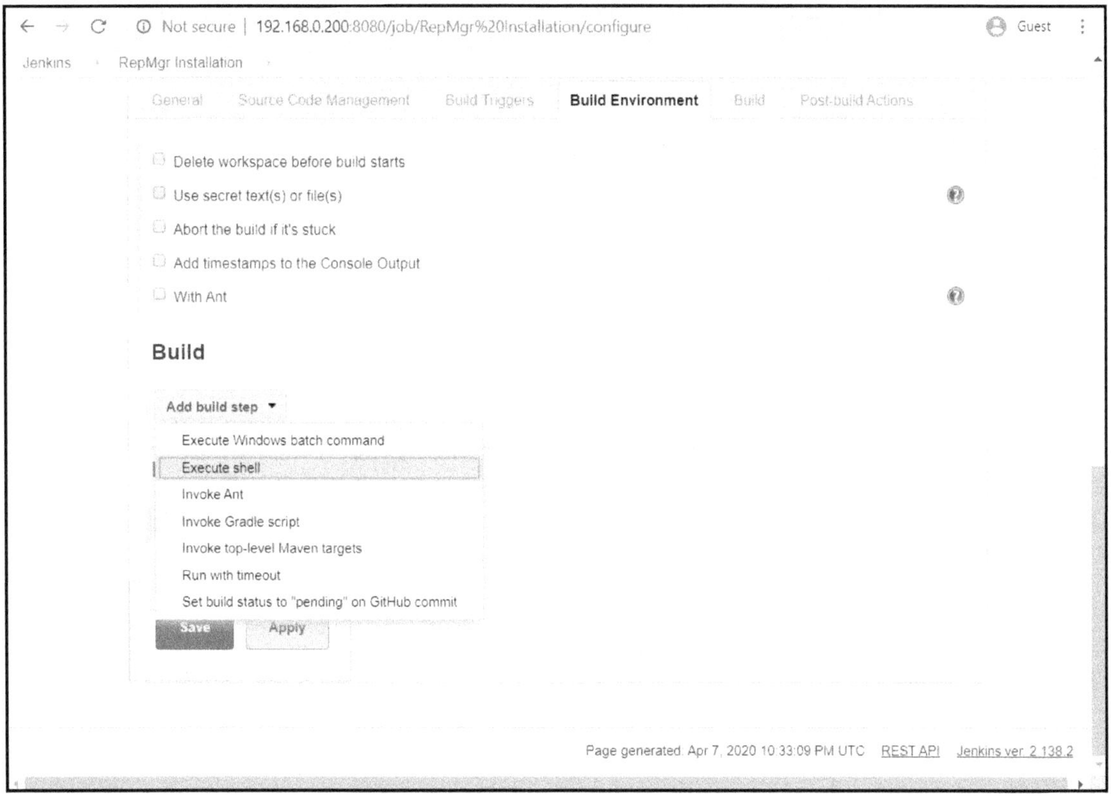

Figure 7.26 – Setting up the Jenkins project to run a shell script

11. Paste the following code into the shell:

```
ssh -o "StrictHostKeyChecking no" vagrant@192.168.0.191 'sudo apt
update -y'
ssh vagrant@192.168.0.191 'sudo apt install -y postgresql-12-
repmgr'
```

12. Then, click **Apply** followed by **Save**:

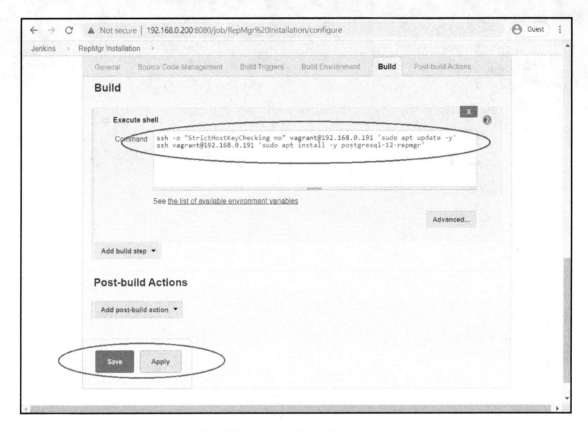

Figure 7.27 – Ssh script to install the RepMgr tool remotely

This Jenkins script will call two remote commands to the PostgreSQL server, `192.168.0.191`, through the `vagrant` user to proceed to `apt update` and to install the `repmgr` tool.

13. Then, click **Build Now**:

Figure 7.28 – Executing our Jenkins project

14. When you see the **#1** build happening with a green color, this means that your Jenkins project has been successful. Click on the **#1** build to drop it down:

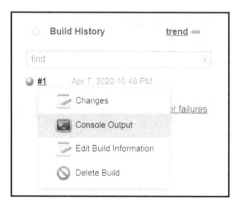

Figure 7.29 – Checking the output of the successful build

15. Select the **Console Output** option:

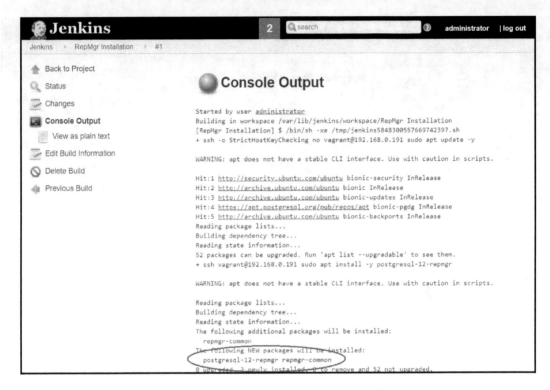

Figure 7.30 – Console Output

16. Click **Log out** to complete your Jenkins session.

17. Now, if you access Vagrant through `ssh` in the PostgreSQL VM to check whether `repmgr` has been installed, you will see it in its entirety:

```
vagrant@vagrant:~$ repmgr --version
repmgr 5.0.0
```

Finally, the output will appear as shown in the following screenshot:

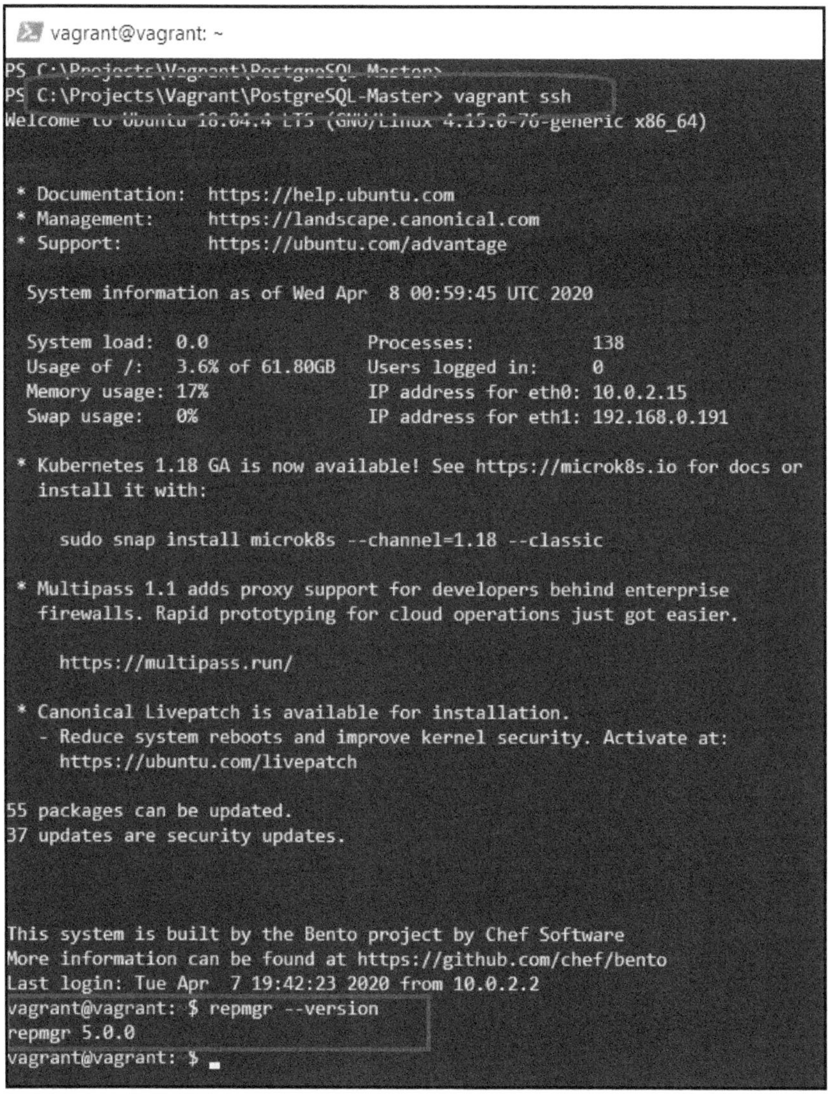

Figure 7.31 – Checking from the PostgreSQL VM to see that RepMgr has been installed remotely

By now, we have deployed PostgreSQL's RepMgr tool by using Jenkins and we've seen how easy it is to implement integration and automated tasks through this excellent tool.

In our everyday working lives, we will use Jenkins to automate tasks on which humans spend too much time; for example, when we are going to implement a new feature and we need to perform tests, let those repetitive tasks be handled by Jenkins. In other words, let an automation system such as Jenkins take charge of building the software, or deploying it on servers, but that this is only allowed when the software has been properly tested and has the green light of approval for it.

Now, we will proceed to use Jenkins with Ansible in the next section.

Creating an Ansible playbook for PostgreSQL

Ansible is the simplest IT automation engine that automates infrastructure, applications, and many other needs. Ansible uses **YAML (Ain't Markup Language)** – a very simple language, to establish Ansible playbooks. An Ansible playbook is composed of one or more 'plays' in an ordered list. The terms 'playbook' and 'play' are sports analogies. Each play executes part of the overall goal of the playbook, running one or more tasks. Each task calls an Ansible module.

We are converting the Jenkins server to be an Ansible server as well. Hence, we can create a Jenkins project named *Ansible Installation*:

1. This is the Jenkins script to install Ansible:

```
sudo apt update -y
echo Y | sudo apt install -y ansible
# set the remote PostgreSQL IP 192.168.0.191
sudo sh -c 'sudo echo "[postgres]" > /etc/ansible/hosts'
sudo sh -c 'sudo echo "192.168.0.191 ansible_user=vagrant
ansible_ssh_private_key_file=/var/lib/jenkins/.ssh/id_rsa" >>
/etc/ansible/hosts'
```

This script firstly makes an `apt update` statement and calls `apt install` for the Ansible tool. Then, it defines a group of Ansible hosts named `postgres`. The `postgres` group now includes our PostgreSQL server, `192.168.0.191`, which Ansible can access through the Vagrant user by means of the `ssh` key file – `/var/lib/jenkins/.ssh/id_rsa`.

2. Now, make a freestyle project called **Ansible Installation** as follows:

Figure 7.32 – Creating a new freestyle project to install Ansible

In the following screenshot, you can see what the **Build** script of the Ansible Installation project looks like:

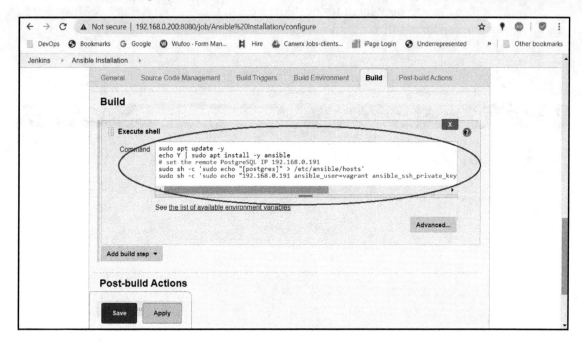

Figure 7.33 – The script inside the Ansible Installation project

3. Once the Ansible Installation project has been built, if we log in to the Jenkins server with `ssh`, we will see that our Ansible installation was successful:

```
vagrant@devopsubuntu1804:~$ ansible --version
ansible 2.9.6
    config file = /etc/ansible/ansible.cfg
    configured module search path =
[u'/home/vagrant/.ansible/plugins/modules',
u'/usr/share/ansible/plugins/modules']
    ansible python module location = /usr/lib/python2.7/dist-
packages/ansible
    executable location = /usr/bin/ansible
    python version = 2.7.15+ (default, Oct 7 2019, 17:39:04) [GCC
7.4.0]

vagrant@devopsubuntu1804:~$ cat /etc/ansible/hosts
[postgres]
192.168.0.191 ansible_user=vagrant
ansible_ssh_private_key_file=/var/lib/jenkins/.ssh/id_rsa
```

Here is the screenshot of the output of the preceding script:

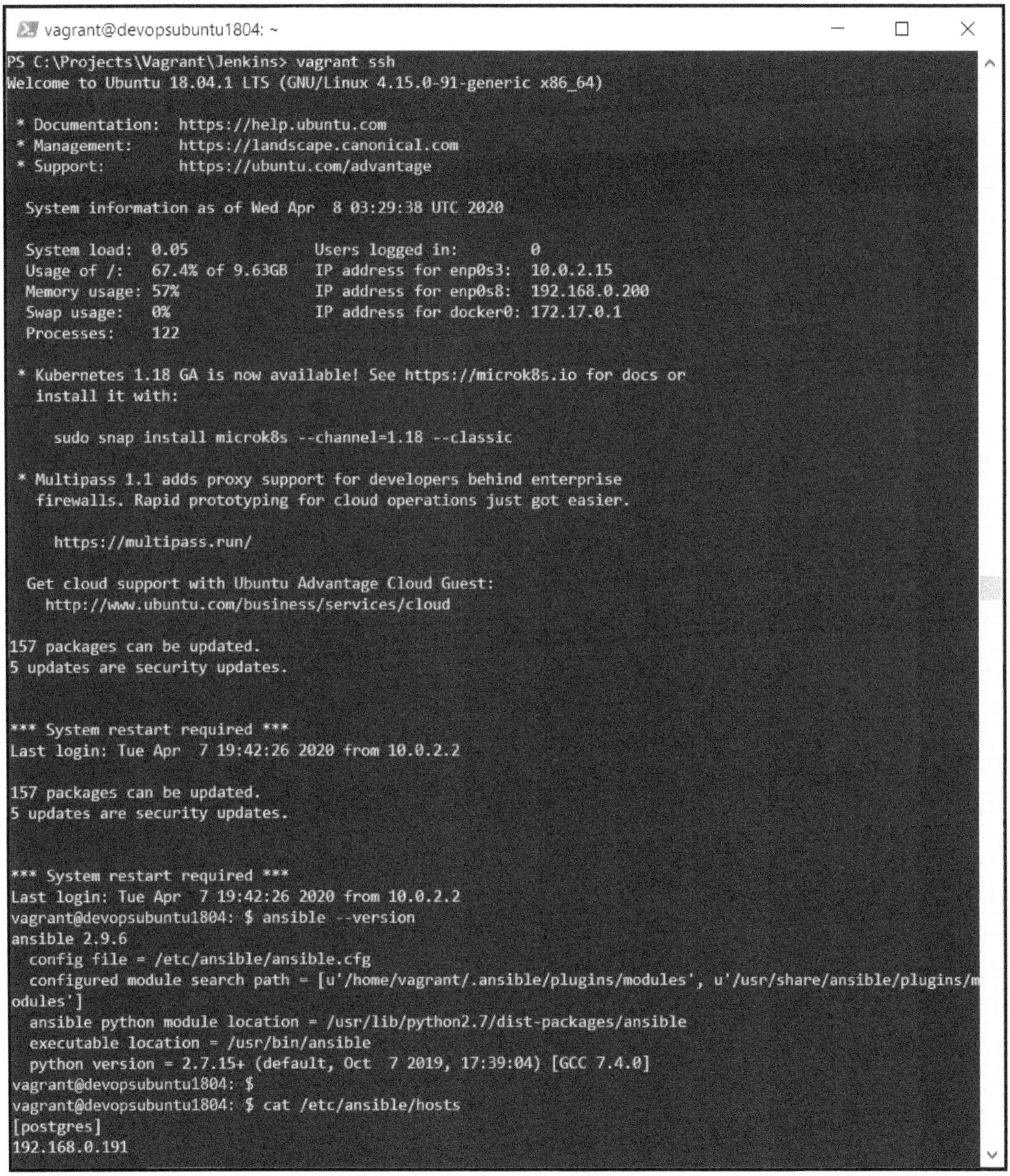

```
vagrant@devopsubuntu1804: ~                                          —   □   ×

PS C:\Projects\Vagrant\Jenkins> vagrant ssh
Welcome to Ubuntu 18.04.1 LTS (GNU/Linux 4.15.0-91-generic x86_64)

 * Documentation:  https://help.ubuntu.com
 * Management:     https://landscape.canonical.com
 * Support:        https://ubuntu.com/advantage

  System information as of Wed Apr  8 03:29:38 UTC 2020

  System load:  0.05               Users logged in:          0
  Usage of /:   67.4% of 9.63GB    IP address for enp0s3:    10.0.2.15
  Memory usage: 57%                IP address for enp0s8:    192.168.0.200
  Swap usage:   0%                 IP address for docker0:   172.17.0.1
  Processes:    122

 * Kubernetes 1.18 GA is now available! See https://microk8s.io for docs or
   install it with:

     sudo snap install microk8s --channel=1.18 --classic

 * Multipass 1.1 adds proxy support for developers behind enterprise
   firewalls. Rapid prototyping for cloud operations just got easier.

     https://multipass.run/

  Get cloud support with Ubuntu Advantage Cloud Guest:
    http://www.ubuntu.com/business/services/cloud

157 packages can be updated.
5 updates are security updates.


*** System restart required ***
Last login: Tue Apr  7 19:42:26 2020 from 10.0.2.2

157 packages can be updated.
5 updates are security updates.


*** System restart required ***
Last login: Tue Apr  7 19:42:26 2020 from 10.0.2.2
vagrant@devopsubuntu1804: $ ansible --version
ansible 2.9.6
  config file = /etc/ansible/ansible.cfg
  configured module search path = [u'/home/vagrant/.ansible/plugins/modules', u'/usr/share/ansible/plugins/m
odules']
  ansible python module location = /usr/lib/python2.7/dist-packages/ansible
  executable location = /usr/bin/ansible
  python version = 2.7.15+ (default, Oct  7 2019, 17:39:04) [GCC 7.4.0]
vagrant@devopsubuntu1804: $
vagrant@devopsubuntu1804: $ cat /etc/ansible/hosts
[postgres]
192.168.0.191
```

Figure 7.34 – Ansible server

4. Create a new Ansible playbook, `/usr/local/src/postgresdata.yml`, on the Jenkins server.

This playbook will download SQL files from GitHub to execute remotely from the Ansible server to the PostgreSQL server:

```
vagrant@devopsubuntu1804:~$ sudo vi /usr/local/src/postgresdata.yml
- name: PostgreSQL Data Playbook
  hosts: postgres
  become: yes
  tasks:
  - name: install prerequisites
    apt: name={{ item }} update_cache=yes state=present
    with_items:
      - python-psycopg2
      - python-ipaddress
  - name: Grant user postgres md5
    postgresql_pg_hba:
      dest: /etc/postgresql/12/main/pg_hba.conf
      contype: local
      users: postgres
      databases: atm
      method: trust
      create: true
  - name: Stop service postgresql, if running
    service:
      name: postgresql
      state: stopped
  - name: Start service postgresql, if not running
    service:
      name: postgresql
      state: started
  - name: download atm table definition
    get_url:
url=https://github.com/lequanha/PostgreSQL-12-Development-and-Admin
istration-Projects/blob/master/Chapter%202/atmdefs.sql?raw=true
      dest=/tmp/atmdefs.sql
  - name: Run queries from atm definition SQL script
    postgresql_query:
      db: atm
      path_to_script: /tmp/atmdefs.sql
      positional_args:
      - 1
  - name: download atm table data insertions
    get_url:
url=https://github.com/lequanha/PostgreSQL-12-Development-and-Admin
istration-
Projects/blob/master/Chapter%202/atmlocations.sql?raw=true
```

```
        dest=/tmp/atmlocations.sql
  - name: Run queries from atm data SQL script
    postgresql_query:
      db: atm
      path_to_script: /tmp/atmlocations.sql
      positional_args:
      - 1
```

In the preceding script, the following is observed:

- The preceding Ansible playbook will execute into remote hosts in the Postgres group. Previously, we defined the Postgres group of hosts inside the /etc/ansible/hosts file, including the IP, 192.168.0.191, of our PostgreSQL server.
- This Ansible playbook contains eight tasks with the names *install prerequisites*, *Grant user Postgres md5*, and *Stop service PostgreSQL*, if running. So, the order of task execution will be the same as the list order of the playbook.
- The first task will install two Python extensions required by Ansible to connect to a PostgreSQL database over the network. These are python-psycopg2 and python-ipaddress. Then, the second Ansible task will add one more entry to the remote pg_hba.conf file of the PostgreSQL server to allow a trust connection without a password for the Postgres user on the atm database.
- After changing the pg_hba.conf file of the PostgreSQL server, Ansible will restart (stop and then start) the PostgreSQL service remotely. Then, it will download SQL script files from GitHub and process these SQL scripts to create an atm table and then populate data for that table.

5. The first GitHub file includes the SQL definition of the **ATM locations** table:

```
← → C    🔒 github.com/lequanha/PostgreSQL-12-Development-and-Admi...        👤 Guest    ⋮

33 lines (26 sloc)   1.01 KB

 1    CREATE SEQUENCE public."ATM locations_ID_seq"
 2        INCREMENT 1
 3        START 658
 4        MINVALUE 1
 5        MAXVALUE 2147483647
 6        CACHE 1;
 7
 8    ALTER SEQUENCE public."ATM locations_ID_seq"
 9        OWNER TO dba;
10
11    GRANT ALL ON SEQUENCE public."ATM locations_ID_seq" TO dba;
12
13    CREATE TABLE public."ATM locations"
14    (
15        "ID" integer NOT NULL DEFAULT nextval('"ATM locations_ID_seq"'::regclass),
16        "BankName" character varying(60) COLLATE pg_catalog."default" NOT NULL,
17        "Address" character varying(50) COLLATE pg_catalog."default" NOT NULL,
18        "County" character varying(15) COLLATE pg_catalog."default" NOT NULL,
19        "City" character varying(15) COLLATE pg_catalog."default" NOT NULL,
20        "State" character(2) COLLATE pg_catalog."default" NOT NULL,
21        "ZipCode" integer NOT NULL,
22        CONSTRAINT "ATM locations_pkey" PRIMARY KEY ("ID")
23    )
24
25    TABLESPACE pg_default;
26
27    ALTER TABLE public."ATM locations"
28        OWNER to dba;
29
```

Figure 7.35 – The ATM locations table definition on GitHub

6. The second GitHub file includes the SQL insert statements for all data of the
   ATM locations table:

```
https://raw.githubusercontent.c    ×    +                                    —    □    ×

←   →   C    🔒 raw.githubusercontent.com/lequanha/PostgreSQL-12-Development-a...    👤 Guest    ⋮

INSERT INTO  public."ATM locations"("BankName","Address","County","City","State","ZipCode")
VALUES('Wells Fargo ATM','500 W 30 STREET','New York','New York','NY',1000),
('JPMorgan Chase Bank, National Association','1260 Broadway','New York','New York','NY',10001),
('Sterling National Bank of New York','1261 Fifth Avenue','New York','New York','NY',10001),
('Bank of America N.A. GA1-006-15-40','1293 Broadway','New York','New York','NY',10001),
('Bank of Hope','16 West 32nd Street','New York','New York','NY',10001),
('TD Bank N.A.','200 West 26th Street','New York','New York','NY',10001),
('Citibank N. A.','201 West 34th Street','New York','New York','NY',10001),
('Capital One, N.A.','215 West 34th Street','New York','New York','NY',10001),
('Citibank N. A.','22 West 32nd Street','New York','New York','NY',10001),
('Sterling National Bank of New York','227 West 27th Street','New York','New York','NY',10001),
('JPMorgan Chase Bank, National Association','245 Seventh Avenue','New York','New
York','NY',10001),
('Amalgamated Bank','275 Seventh Avenue','New York','New York','NY',10001),
('JPMorgan Chase Bank, National Association','305 Seventh Avenue','New York','New
York','NY',10001),
('Woori America Bank','330 Fifth Avenue','New York','New York','NY',10001),
('Commerce Bank, N.A.','341 Ninth Avenue','New York','New York','NY',10001),
('TD Bank N.A.','350 West 31st Street','New York','New York','NY',10001),
('USAlliance Financial','350 West 31st Street','New York','New York','NY',10001),
('Bank of America N.A. GA1-006-15-40','358 Fifth Avenue','New York','New York','NY',10001),
('Sterling National Bank of New York','406 West 31st Street','New York','New York','NY',10001),
('Sterling National Bank of New York','7 Penn Plaza','New York','New York','NY',10001),
('Bank of America N.A. GA1-006-15-40','800 Avenue of the Americas','New York','New
York','NY',10001),
('HSBC Bank USA, National Association','800 Avenue of the Americas','New York','New
York','NY',10001),
('TD Bank N.A.','885 Sixth Avenue','New York','New York','NY',10001),
('BankUnited, NA','960 Avenue of the Americas','New York','New York','NY',10001),
('TD Bank N.A.','Five Penn Plaza','New York','New York','NY',10001),
('JPMorgan Chase Bank, National Association','109 Delancy Street','New York','New
York','NY',10002),
('First American International Bank','123 EAST BROADWAY','New York','New York','NY',10002),
('Banco Popular North America','134 Delancy Street','New York','New York','NY',10002),
('Cathay Bank','16-18 East Broadway','New York','New York','NY',10002),
('First American International Bank','240 Grand Street','New York','New York','NY',10002),
('First American International Bank','29 Bowery','New York','New York','NY',10002),
('HSBC Bank USA, National Association','307 Grand Street','New York','New York','NY',10002),
('Banco Popular North America','310 Houston Street','New York','New York','NY',10002),
('TD Bank N.A.','314 Grand Street','New York','New York','NY',10002),
```

Figure 7.36 – Data of the ATM locations table on GitHub

7. Copy these statements to execute the Ansible playbook by using the `ansible-playbook` command:

```
vagrant@devopsubuntu1804:~$ sudo su
root@devopsubuntu1804:/home/vagrant# su jenkins
jenkins@devopsubuntu1804:/home/vagrant$ ansible-playbook
/usr/local/src/postgresdata.yml
```

You should get no errors and the output should look like the following screenshot:

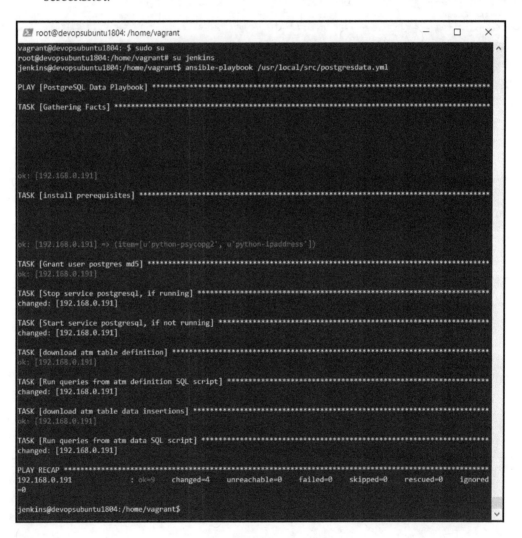

Figure 7.37 – Execution of an Ansible playbook

8. Once the execution of Ansible is complete, switch to the PostgreSQL server to check whether a new table containing full data, ATM locations, has been created:

```
vagrant@vagrant:~$ sudo su
root@vagrant:/home/vagrant# su postgres
postgres@vagrant:/home/vagrant$ psql
postgres=# \c atm
atm=# \d "ATM locations"
```

As shown in the following screenshot, we are able to list the ATM locations table and also retrieve the data by SQL:

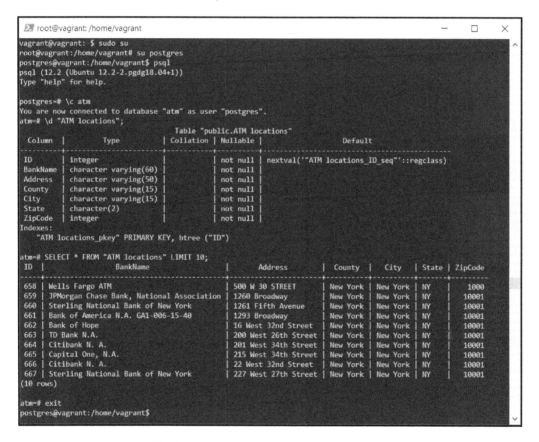

Figure 7.38 – The Ansible playbook has created the ATM locations table with data remotely

In this section, we looked at the Ansible playbook to create a PostgreSQL table and also to populate ARM data for the table. We are now moving to the next section about Terraform, a very similar DevOps tool to Ansible.

Managing PostgreSQL by Terraform

Even though Ansible and Terraform both offer configuration automation, the main difference between Ansible and Terraform is that Terraform provides reliable orchestration, while Ansible provides robust configuration management. This means that Ansible will be good for installing and managing software on existing servers while Terraform is more focused on cloud and infrastructure tasks such as load balancers, databases, or networking:

1. To launch a new Terraform server, use this Vagrant box, `servetguney/ubuntu-18-04-terraform-git`, here: `https://app.vagrantup.com/servetguney/boxes/ubuntu-18-04-terraform-git`.

2. However, Terraform installation is straightforward, so we now proceed to install Terraform on our Jenkins server as well.

    Check the following home page for the latest Terraform version for Ubuntu 18: `https://www.terraform.io/downloads.html`.

    Then we download it into the Jenkins server, decompress the file, and then Terraform can be used right away by copying the Terraform file into the `bin` folder of the VM:

    ```
    vagrant@devopsubuntu1804:~$ sudo su
    root@devopsubuntu1804:/home/vagrant# cd /tmp
    root@devopsubuntu1804:/tmp# wget
    https://releases.hashicorp.com/terraform/0.12.24/terraform_0.12.24_
    linux_amd64.zip

    root@devopsubuntu1804:/tmp# apt install unzip
    root@devopsubuntu1804:/tmp# unzip terraform_0.12.24_linux_amd64.zip
    root@devopsubuntu1804:/tmp# mv terraform /usr/local/bin/
    root@devopsubuntu1804:/tmp# terraform --version
    Terraform v0.12.24
    ```

3. Use the PostgreSQL provider of Terraform to create a sample database inside the current PostgreSQL server: `/usr/local/src/terraform/main.tf`:

    ```
    vagrant@devopsubuntu1804:~$ sudo mkdir /usr/local/src/terraform
    vagrant@devopsubuntu1804:~$ sudo vi
    /usr/local/src/terraform/main.tf
    --------------------------------------------
    provider "postgresql" {
        host = "192.168.0.191"
        port = 5432
        username = "postgres"
    ```

```
        sslmode = "require"
        connect_timeout = 15
    }
    resource "postgresql_role" "terraform_role" {
        name = "terraform_role"
        login = true
        password = "tfdemo"
    }
    resource "postgresql_role" "terraform_dba" {
        name = "terraform_dba"
        login = true
        password = "tfdba"
    }
    resource "postgresql_database" "terraform_db" {
        name = "terraform_db"
        owner = "terraform_dba"
        lc_collate = "C"
        connection_limit = -1
        allow_connections = true
    }
    resource "postgresql_schema" "terraform_schema" {
        name = "terraform_schema"
        owner = "postgres"
        database = postgresql_database.terraform_db.name
        # terraform_role can create new objects in the schema. This is
the role that
        # migrations are executed as.
        policy {
            create = true
            usage = true
            role =  postgresql_role.terraform_role.name
        }
        policy {
            create_with_grant = true
            usage_with_grant = true
            role =  postgresql_role.terraform_dba.name
        }
    }
}
------------------------------------------------
```

This Terraform template will connect to the PostgreSQL server, `192.168.0.191`, on port `5432`. After connecting to the remote PostgreSQL server, Terraform creates two users: a username, `terraform_role`, with the password `tfdemo`, and a second username, `terraform_dba`, with the password `tfdba`. Through the user `terraform_dba`, Terraform creates a PostgreSQL database named `terraform_db`.

Inside this database, Terraform also creates a schema named `terraform_schema` and grants `CREATE` and `USAGE` policies for the `terraform_role` user on this schema (the other `terraform_dba` user has to CREATE privileges, the ability to GRANT the `CREATE` privilege, `USAGE` privileges and the ability to GRANT the `USAGE` privilege for other database roles).

4. In order to connect Terraform successfully from the Jenkins server, `192.168.0.200`, to the PostgreSQL server, we need to adjust the PostgreSQL `pg_hba.conf` file inside the other PostgreSQL server, `192.168.0.191`:

```
vagrant@vagrant:~$ sudo su
root@vagrant:/home/vagrant# su postgres
postgres@vagrant:/home/vagrant$ vi
/etc/postgresql/12/main/pg_hba.conf
-----------------------------------------------
...
# host all postgres 0.0.0.0/0 reject
host all postgres 192.168.0.200/32 trust
...
-----------------------------------------------
```

Turn off the rejection rules such as `host all postgres 0.0.0.0/0 reject` and then add a trust rule for the Jenkins server to connect through a Postgres user without the need for a password.

5. After that, restart the service on the PostgreSQL server, `192.168.0.191`:

```
postgres@vagrant:/home/vagrant$ exit
root@vagrant:/home/vagrant# service postgresql stop
root@vagrant:/home/vagrant# service postgresql start
```

6. Now we come back to the Jenkins server. Go to the `main.tf` folder:

```
vagrant@devopsubuntu1804:~$ cd /usr/local/src/terraform
vagrant@devopsubuntu1804:/usr/local/src/terraform$ sudo terraform
init
vagrant@devopsubuntu1804:/usr/local/src/terraform$ sudo terraform
plan
vagrant@devopsubuntu1804:/usr/local/src/terraform$ echo yes | sudo
terraform apply
```

The `terraform init` command is used to initialize a working directory containing Terraform configuration files. Then, we create an execution plan using the `terraform plan` command. Finally, we execute the Terraform template by using the `terraform apply` command.

You should get no errors and the output should be as shown in the following screenshot:

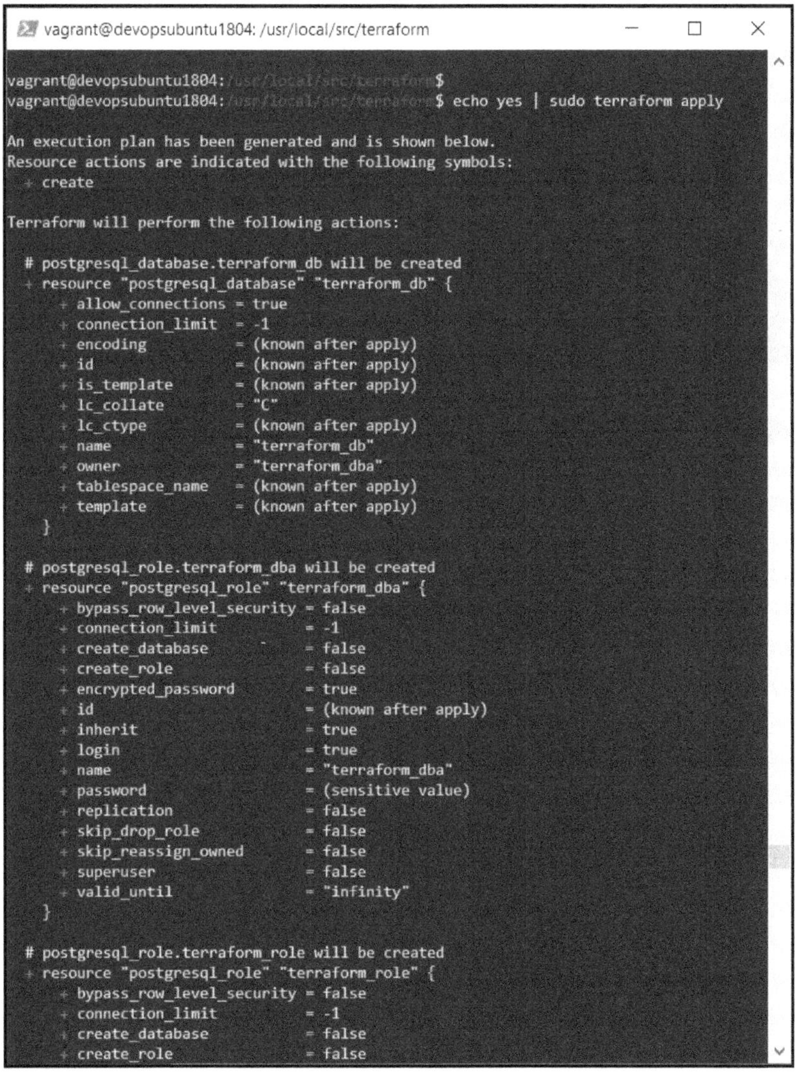

Figure 7.39 – Terraform application of the main.tf file

7. We can now come back to the PostgreSQL server to perform a check:

```
root@vagrant:/home/vagrant# su postgres
postgres@vagrant:/home/vagrant$ psql
postgres=# \l
```

We will see the created `terraform_db` database inside the result:

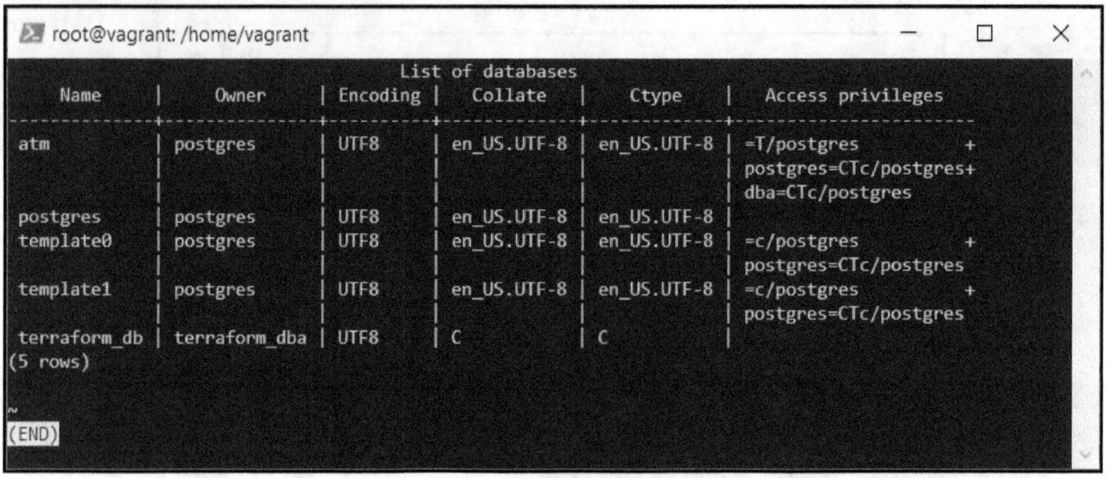

Figure 7.40 – The new terraform_db database has been created

8. We will be able to see our new Terraform created roles and our new schema that has been created:

```
postgres=# \c terraform_db
You are now connected to database "terraform_db" as user
"postgres".
terraform_db=# \dn
 List of schemas
 Name               | Owner
--------------------+----------
 public             | postgres
 terraform_schema   | postgres
 (2 rows)

terraform_db=# \du
 List of roles
 Role name          | Attributes
| Member of
----------------+-------------------------------------------------
-----------+------------
```

```
    dba             |
 | {}
    postgres        | Superuser, Create role, Create DB, Replication,
 Bypass RLS | {}
    terraform_dba   | Password valid until infinity
 | {}
    terraform_role  | Password valid until infinity
 | {}
```

This section has shown you how to create a Terraform template to automate the creation of a new PostgreSQL schema, a new PostgreSQL database, and new roles.

Summary

In this chapter, we learned about some good tools for DevOps to implement a single solution with PostgreSQL step by step. So far, system administrators have seen VirtualBox, Vagrant, and Puppet launch PostgreSQL in VMs. They can also apply Jenkins scripts, Ansible playbooks, and Terraform templates to deploy PostgreSQL servers, PostgreSQL schemas, databases, roles, and tables automatically.

In the next chapter, we will focus on PostgreSQL high availability, so if you are interested to see how our database is always available, even in the event of a disaster, I invite you to read the next chapter.

8
PostgreSQL High Availability Clusters

In this chapter, we will learn about **High Availability** (**HA**) clusters and how to achieve them with PostgreSQL. HA database clusters (also known as failover clusters) are a group of computers that support server applications that can be reliably utilized with a minimum of downtime in case something happens, such as a disaster, or we simply need to carry out an upgrade of the hardware or software.

As we have seen in the preceding definition, when we talk about high availability, this will always mean more than two servers, where one of them functions as the master and another one as the standby. This kind of architecture is the minimum that will be needed to comply with zero or near-zero downtime.

In the PostgreSQL world, it is common for architectures such as HA to use streaming replication as a base. Failover is managed is through an application that constantly verifies that both servers are synchronized and up and running. This application will automatically ensure that when, for example, the master has a failure, the requests are transferred to the standby server, thereby achieving the much sought-after high availability.

In the following topics, we will address and show how to achieve this. You will learn the following:

- Setting up streaming replication on PostgreSQL
- Setting up a PostgreSQL HA cluster through the Heimdall data proxy

Technical requirements

This chapter will take developers around 24-36 work hours to develop a typical load balancer and throughputs.

Setting up streaming replication on PostgreSQL

As we saw in the introduction to this chapter, streaming replication is commonly used as the basis for a high availability architecture because we have our instance that is replicated in one or more servers.

PostgreSQL has two types of replication: synchronous and asynchronous. In synchronous replication, a COMMIT is only valid if it was confirmed by all the PostgreSQL servers used, which guarantees that the data will never be lost.
In asynchronous replication, data can arrive at the standby server AFTER the transaction has been committed on the master server. In many cases, this is fully accepted because asynchronous replication promises little overhead and does not impact the performance of the primary server, which is why asynchronous replication is the standard method of replication and will be developed further throughout this topic.

In the previous chapter, we created our first Ubuntu server with PostgreSQL and this will function as our master. The following steps were performed in the preceding chapter and will help us to create our standby server. It is worth reviewing them again, so let's see how:

```
PS C:\Windows\system32> mkdir C:\Projects\Vagrant\PostgreSQL-Standby
PS C:\Windows\system32> cd C:\Projects\Vagrant\PostgreSQL-Standby\
PS C:\Projects\Vagrant\PostgreSQL-Standby> bcdedit /set
hypervisorlaunchtype off
PS C:\Projects\Vagrant\PostgreSQL-Standby> vagrant init bento/ubuntu-18.04
```

Following the preceding steps, we will edit the Vagrant file
at C:\Projects\Vagrant\PostgreSQL-Standby\Vagrantfile and will then add
PostgreSQL installation lines as follows:

```
Vagrant.configure("2") do |config|
    config.vm.box = "bento/ubuntu-18.04"
    config.vm.network "public_network", ip: "192.168.0.192"
    config.vm.provision "shell", inline: <<-SHELL
        sh -c 'echo "deb http://apt.postgresql.org/pub/repos/apt
$(lsb_release -s)-pgdg main" > /etc/apt/sources.list.d/pgdg.list'
        wget --quiet -O -
https://www.postgresql.org/media/keys/ACCC4CF8.asc | sudo apt-key add -
        apt-get update
        apt-get -y install postgresql-12
    SHELL
 end
```

With the previous configuration, we will start up our standby PostgreSQL server:

```
C:\Projects\Vagrant\PostgreSQL-Standby> vagrant up --provider virtualbox
```

When we watch on VirtualBox Manager, we can see our master (created previously) and our standby servers:

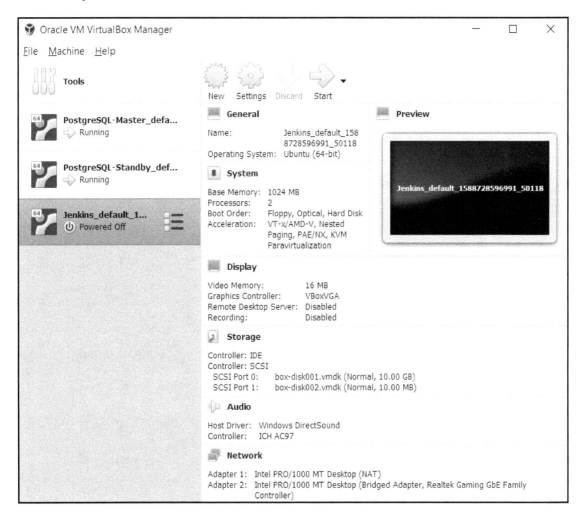

Figure 8.1 Virtualbox homepage

As you can see, both servers are running and we will now configure our standby server in order to work as a replica. Let's see how to do this.

First, there are four things we have to do on the master server:

1. Enable networking in `postgresql.conf`, where the `listen_addresses` parameter should have `*`. We will see what this looks like:

```
# vagrant ssh
vagrant@vagrant:~$ sudo su postgres
postgres@vagrant:~$ psql
postgres=# SHOW listen_addresses;
 listen_addresses
-------------------
 *
```

2. A good habit, but something that isn't mandatory, is to create a replica user:

```
postgres=# CREATE USER replicator REPLICATION;
CREATE ROLE
```

3. Allow remote access for replication work in the `pg_hba.conf` file. Just for demonstration purposes, we will use *trust* as authentication. Please do not use this in production environments:

```
host replication replicator 192.168.0.192/32 trust # Stanby ip
addresses
```

4. Restart PostgreSQL:

```
vagrant@vagrant:~$ sudo /etc/init.d/postgresql restart
[ ok ] Restarting postgresql (via systemctl): postgresql.service.
```

Now, it's time to work on the standby server, which, as we said previously, will function as a replica. Perform the following steps:

1. In the standby server, stop PostgreSQL:

```
#vagrant ssh
vagrant@vagrant:~$ sudo /etc/init.d/postgresql stop
[ ok ] Stopping postgresql (via systemctl): postgresql.service.
```

2. Empty the data directory:

```
vagrant@vagrant:~$ sudo su postgres
postgres@vagrant:~$ cd /var/lib/postgresql/12/main/
postgres@vagrant:~/12/main$ ls
backup_label.old pg_dynshmem pg_replslot pg_stat_tmp PG_VERSION
postmaster.opts
base pg_logical pg_serial pg_subtrans pg_wal standby.signal
global pg_multixact pg_snapshots pg_tblspc pg_xact
```

```
pg_commit_ts pg_notify pg_stat pg_twophase postgresql.auto.conf
postgres@vagrant:~/12/main$ rm -rf *
```

3. Create a base backup with `pg_basebackup`. More information can be found at https://www.postgresql.org/docs/12/app-pgbasebackup.html. Basically, what it does is make a complete binary copy of a PostgreSQL instance. This command does its job without affecting other clients in the database and can be used both for point-in-time recovery and as the starting point for log shipping or streaming replication standby servers. Let's see how this works:

```
postgres@vagrant:~/12/main$ pg_basebackup -h 192.168.0.191 -D
/var/lib/postgresql/12/main -P -U replicator -Xs -R --
checkpoint=fast
```

As you can see, there are a couple of parameters:

- `-h`: Master server IP addresses.
- `-D`: The directory to write the output to.
- `-P`: Indicates progress.
- `-U`: User.
- `-checkpoint`: We set fast speeds in the checkpoint process so the copy can begin.
- `-Xs`: Streams the write-ahead log while the backup is created.
- `-R`: Creates a `standby.signal` file to facilitate setting up a standby server.

When `pg_basebackup` is complete, we will check two files – `standby.signal` and `postgresql.auto.conf`. Both are created thanks to the `-R` parameter. Now we will see what they look like:

```
postgres@vagrant:~/12/main$ ls standby.signal -l
-rw------- 1 postgres postgres 0 May 15 19:30 standby.signal

postgres@vagrant:~/12/main$ cat postgresql.auto.conf
# Do not edit this file manually!
# It will be overwritten by the ALTER SYSTEM command.
primary_conninfo = 'user=postgres
passfile=''/var/lib/postgresql/.pgpass'' channel_binding=prefer
host=192.168.0.191 port=5432 sslmode=prefer sslcompression=0
ssl_min_protocol_version=TLSv1.2 gssencmode=prefer
krbsrvname=postgres target_session_attrs=any'
```

Now, we are ready to start our standby server. Execute the following commands:

```
vagrant@vagrant:~$ sudo /etc/init.d/postgresql start
[ ok ] Starting postgresql (via systemctl): postgresql.service.

vagrant@vagrant:~$ ps -ef |grep postgres
postgres 1052 1 0 21:32 ? 00:00:01 /usr/lib/postgresql/12/bin/postgres -D
/var/lib/postgresql/12/main -c
config_file=/etc/postgresql/12/main/postgresql.conf
postgres 1053 1052 0 21:32 ? 00:00:00 postgres: 12/main: startup recovering
000000010000000000000010
postgres 1195 1052 0 21:32 ? 00:00:00 postgres: 12/main: checkpointer
postgres 1196 1052 0 21:32 ? 00:00:00 postgres: 12/main: background writer
postgres 1202 1052 0 21:32 ? 00:00:00 postgres: 12/main: stats collector
postgres 1203 1052 0 21:32 ? 00:00:04 postgres: 12/main: walreceiver
streaming 0/107D9038
```

As you can see, our standby server is functioning and the `walreceiver` process is working. From now on, every piece of data that is written on the **write-ahead log** (**wal**) master server will be replicated on the standby server. In this way, we have one part of our "failover system," and it is not an automatic process. Why do we say this? Because, in the event that something does go wrong, our master server breaks, for example, and it is unrecoverable, someone will have to promote the standby server to the primary role.

Just by way of an example, we show the command to use in the standby server in case we need to promote it to the role of master:

```
/usr/lib/postgresql/12/bin/pg_ctl promote -D /var/lib/postgresql/12/main
```

So far, we can seen how to have an exact and secure replica of PostgreSQL through streaming replication. In the next topic, we will see how to accomplish high availability automatically through a Heimdall data proxy.

Setting up a PostgreSQL HA cluster through the Heimdall data proxy

Previously, we saw how to create a simple and strong fallback architecture through streaming replication in PostgreSQL, but where human intervention is necessary in case things go wrong.

To get around this issue, automation is key. That is why we are now going to see one of the most interesting solutions in the field of HA: the Heimdall data proxy.

The Heimdall proxy is a data access layer for application developers, database administrators, and architects. Whether on-premises or in the cloud, our proxy helps organizations to deliver faster, more reliable, and secure content generation.

Heimdall has the following features:

- Connection pooling
- Automated caching
- Load balancing
- High availability

In particular, we will talk about HA where Heimdall provides a multi-tiered approach to supporting it and supports a wider variety of database vendors and database topologies. We will examine one of the most simple ways to configure HA through a friendly interface and, more importantly, it is rock-solid. Let's see each step that we will be developing in this topic:

- Heimdall installation by Docker
- Heimdall Webapp and Wizard configuration
- Testing load balancing and HA

Heimdall installation by Docker

Heimdall is available in the cloud from AWS or we can install it on-premises. For practical purposes, here, we will install it locally through Docker by means of the following simple steps:

1. Download the Docker file using the `wget` command:

   ```
   wget https://s3.amazonaws.com/s3.heimdalldata.com/Dockerfile
   ```

2. Build a Docker image:

   ```
   docker build -t "heimdall:current" .
   ```

3. Create a container based on that image in order to obtain Heimdall data:

   ```
   docker run -d --name heimdall-instance -p 8087:8087 -p 3306:3306 -p
   5432:5432 -p 1433:1433 heimdall:current
   ```

So far, the last step shows something super important: -p 8087:8087. This is the port for the Heimdall administration web. Let's see how to configure this first before checking whether our Heimdall container is running:

```
#docker ps -a
CONTAINER ID IMAGE COMMAND CREATED STATUS PORTS NAMES
e96017aef7b5 heimdall:current "/opt/heimdall/heimd..." 1 min ago Up 1 min ago
0.0.0.0:1433->1433/tcp, 0.0.0.0:3306->3306/tcp, 0.0.0.0:5432->5432/tcp,
0.0.0.0:8087->8087/tcp heimdall-instance
```

As you can see, our container is running and it's available for the next steps, where we will set it up.

Heimdall Webapp and Wizard configuration

In order to continue with the configuration, it is necessary to access the following URL, http://localhost:8087, through a browser. The default Heimdall server username is admin and the password is heimdall:

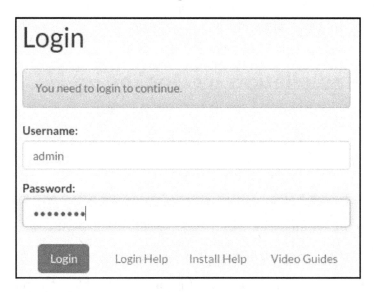

Fig 8.2 Heimdall login

Once the login is successful, we will see the first screen where Heimdall provides us with the following warning, **No virtual databases are configured**, as can be seen in the following screenshot:

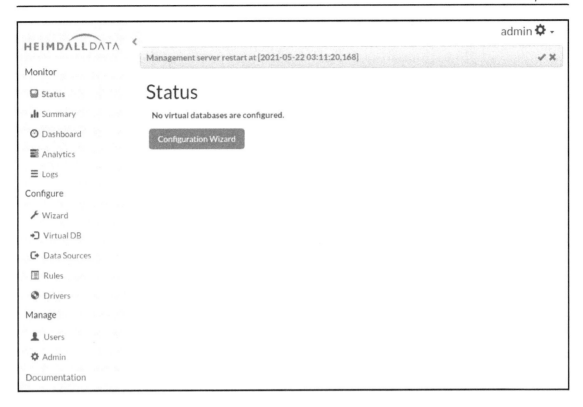

Fig 8.3 Heimdall homepage with the warning

The next step that we must proceed with is **Configuration Wizard**. Click on that button and then let's cover each step in turn:

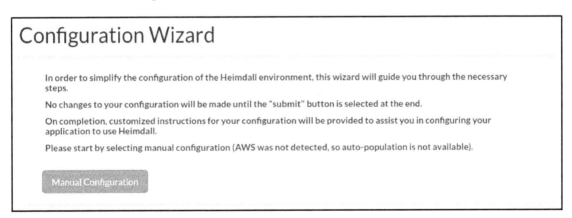

Fig 8.4 Configuration wizard for Heimdall environment

The preceding step, as you can see, is akin to a presentation where it invites us to proceed with the next steps. Click on the **Manual Configuration** button and we will continue with this configuration wizard:

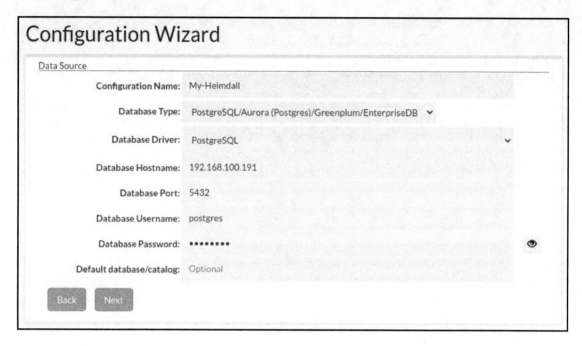

Fig 8.5 Manual configuration details

As you can see in the preceding screenshot, it is necessary to configure the driver, IP addresses, port, and credentials. It's important to mention that the information provided relates to our master server, which was configured in the previous topic. Let's now continue with the next step:

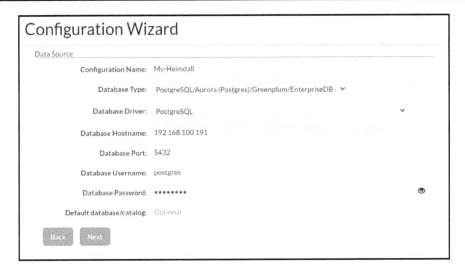

Figure 8.6 Configuration wizard HA changes

You may recall that earlier we said"The most simple way to configure HA". In the preceding screenshot, we have seen how that phrase has been fulfilled because all we have needed to do is to click the **Enable Load Balancing** and **Track Cluster Changes** checkboxes and with that information alone, Heimdall is smart enough to detect our slave or replica server.

That's all that Heimdall needs to know to configure HA, so let's continue with the next step:

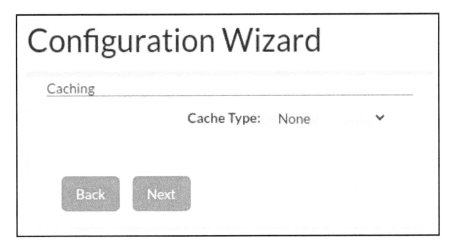

Figure 8.7 Caching changes

In the preceding screenshot, you can see that Heimdall has offered us one interesting option: Caching, which is not covered in this chapter. All we will say about it, for now, is that Heimdall has a two-layer cache system that, from the perspective of performance, should be beneficial in production environments where millions of similar requests are received per second. Now, we will go through the next step of the wizard:

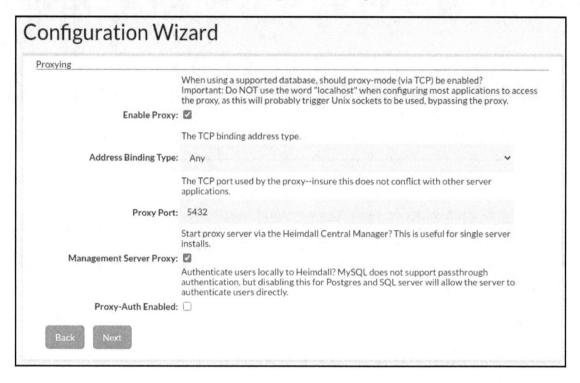

Figure 8.8 Proxy values

So far, we know that Heimdall acts as a proxy and that is exactly what this step offers us. So, at this point, it is important to mention that when, in the future, we connect to our database, this will not be done directly, but through Heimdall. We are almost at the end of the wizard, so let's continue with the following step:

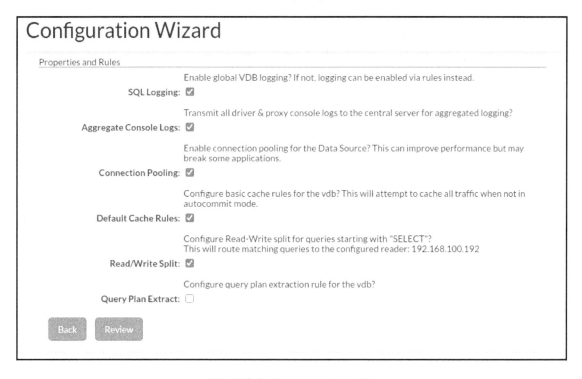

Figure 8.9 Configuration wizard properties and rules

In the preceding screenshot, you can see all the different options available under this step, and one of the most important options from the point of view of performance is **Read/Write Split**. You can see how Heimdall automatically gets the IP addresses of our standby server. In the following step, Heimdall offers a complete review of our configuration, shown as follows:

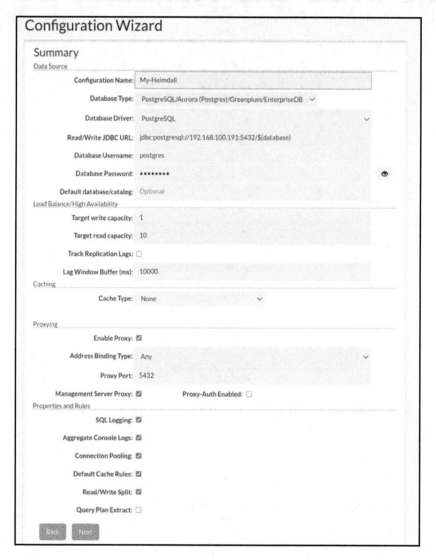

Figure 8.10 Summarization of the configuration wizard

The preceding screenshot is just a summary of all the previous steps, but it will help us to do a review of our configuration. It is also important to know that now, Heimdall will receive all the requests from clients. This means that from the point of view of the client, Heimdall is to the fore and clients will not know whether they are connecting to the master or the standby server. Let's now see how to connect to our new Heimdall cluster by means of psql:

```
# psql -h localhost -p 5432 -d atm -U postgres
atm=>
```

Recall that in our previous chapter, *PostgreSQL with DevOps for Continuous Delivery*, our current master server was created, and in it we created an atm database with Puppet and deployed and populated an ATM locations table through Ansible. Now, in the following topic, we will use this for our high availability test.

Testing load balancing and high availability

Everything done so far has been with the aim of achieving high availability in our database architecture. From now on, when we have an established connection to the database, it should not be lost in the event of a disaster, such as the crashing of our master server. Here, we will instigate a crash in our master server, leaning on Vagrant to suspend it, and we will see how Heimdall acts in this instance.

In order to shake things up a bit, we will run the following query on psql:

```
atm=>
SELECT
    format('select pg_sleep(2), "BankName", "Address" from "ATM locations"
where                         "ID"=%s;', "ID")
FROM
    "ATM locations";
                                      format
---------------------------------------------------------------------------
-----
 select pg_sleep(2), "BankName", "Address" from "ATM locations" where
"ID"=658;
 select pg_sleep(2), "BankName", "Address" from "ATM locations" where
"ID"=659;
 ...
 ...
 select pg_sleep(2), "BankName", "Address" from "ATM locations" where
"ID"=1310;
 select pg_sleep(2), "BankName", "Address" from "ATM locations" where
"ID"=1311;
```

```
(654 rows)

atm => \gexec

 pg_sleep | BankName | Address
----------+-----------------+------------------
          | Wells Fargo ATM | 500 W 30 STREET
(1 row)

 pg_sleep | BankName | Address
----------+-------------------------------------------+---------------
          | JPMorgan Chase Bank, National Association | 1260 Broadway
(1 row)
...
...
```

The preceding query generated several SELECTS (654 queries), as you can see, using the `format` function, and each one of them has a `pg_sleep(2)` function that makes the current session's process sleep until 2 seconds have elapsed, and then executes all the sentences by means of `\gexec`, which is a psql command that takes output and sends it to the server.

As each query will have 2 seconds of artificial delay, the entire process will take around 21 minutes and, at some point during that time, we will suspend our master server with the goal of simulating a crash situation.

First, we will see the current status through the nice interface that Heimdall provides in the **Dashboard** option. Meanwhile let's remember that our queries are running behind schedule:

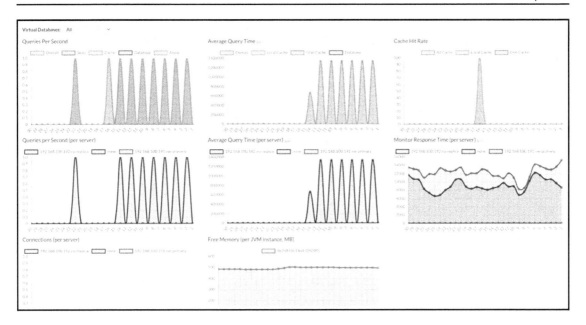

Figure 8.11 Heimdall dashboard

The important thing here is to identify the fact that Heimdall has named our master server as **192.168.100.191-rw-primary**, and the standby server as **192.168.100.192-ro-replica**. At the moment, Heimdall is doing its job as a load balancer and our queries are running, as can be seen in the preceding screenshot. We will now see what happens when we suspend our master server and, at the same time, promote our slave server (replica) through Vagrant. Let's do that now:

```
 # vagrant suspend
==> default: Saving VM state and suspending execution...
```

Following the previous action, we promote our replica:

```
# vagrant ssh -c 'su -c "/usr/lib/postgresql/12/bin/pg_ctl promote -D
/var/lib/postgresql/12/main" postgres'
```

Now we will see what the dashboard looks like once our master is no longer available. Refer to the following screenshot:

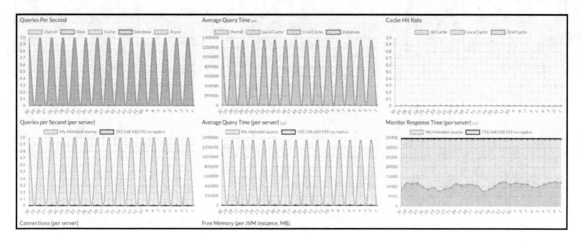

Figure 8.12 Heimdall dashboard without master

As you can see in the preceding screenshot, in the **Monitor Response Time** chart, there are a number of important issues that need to be highlighted:

1. The **192.168.100.191-rw-primary** (master) line in green has gone as a consequence of our artificial crash implemented with `Vagrant suspend`.

2. As a result of the event in point 1, Heimdall acts quickly, changing the role of the **192.168.100.192-ro-replica** line in black and placing at the front the **My-Heimdall-source** green line until a new slave or replica is added.

3. Let's not lose sight of the **Queries per Second** chart. This is perhaps the most important event because it means that end users have never lost their connection to the database.

Following these events, let's now continue monitoring and we will see that everything has returned to normal, but in a new context:

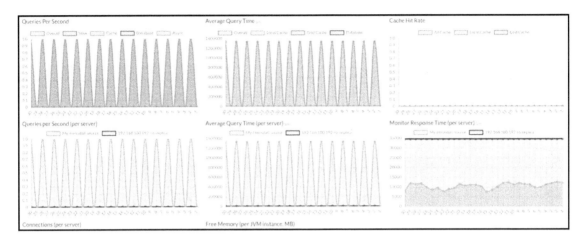

Figure 8.13 Heimdall dashboard post crash

As you can see in the preceding screenshot, in this new context, we can say that **192.168.100.192-ro-replica** is now the one that will receive all the connections and keep the existing ones. This is what is known as high availability, and for end users such as our psql client or web apps, this will be transparent. In the meantime, our technical staff will have taken care of this, rectifying issues on the master server or replacing it with a new one.

Everything that we have seen, and that Heimdall has solved fantastically behind the scenes, is amazing from the perspective of availability, and even more so if we have users concurrently accessing our services.

Summary

In this chapter, we learned about streaming replication with PostgreSQL and saw how straightforward it is to configure it. We stated that it is the base solution for numerous high availability architectures, and then we presented Heimdall, which, based on streaming replication and thanks to its load balancing and high availability features, demonstrated how, in the event of a failure in the master server, it can continue to provide a service without the end customer being aware of this. In the next chapter, we'll explore New Relic and how we can use it to create team dashboards for various applications.

9

High-Performance Team Dashboards Using PostgreSQL and New Relic

In this chapter, you will learn how to install and activate PostgreSQL integration and will gain an understanding of the data collected by the New Relic infrastructure. New Relic PostgreSQL on-host integration receives and sends inventory metrics from our PostgreSQL database to the New Relic platform, where we can aggregate and visualize key performance metrics. Data about the level of instances, databases, and clusters can help us to more easily monitor our PostgreSQL database. You will learn how to use New Relic to monitor a PostgreSQL RDS and you will be able to access, visualize, and troubleshoot entire PostgreSQL databases for yourself – and your software team.

With the help of the project demonstrated in this chapter, we will create a New Relic monitoring dashboard for the PostgreSQL 12 RDS from Amazon Web Services from the previous chapters.

The following topics will be covered in this chapter:

- Signing up for and installing New Relic
- Defining PostgreSQL RDS role permissions
- Configuring New Relic for PostgreSQL
- Adding new metric data for PostgreSQL
- Infrastructure inventory data collection

Technical requirements

This chapter will take developers around 6-8 hours of work to create a New Relic monitoring dashboard for the PostgreSQL 12 RDS. For this chapter, you only need a PostgreSQL RDS and a stable internet connection because New Relic is an online dashboard from the web.

Signing up for and installing New Relic

New Relic supplies a high-performance team dashboard for various systems, such as the following:

- Backend, frontend, and mobile applications
- Cloud and platform technologies
- Host operating systems
- Log ingestion
- Infrastructure
- Open source monitoring systems

We can plug New Relic into different cloud platforms such as **Amazon Web Services** (**AWS**), Microsoft Azure, and Google Cloud Platform. Hence, obviously, New Relic installation for a PostgreSQL RDS from AWS is a typical application of New Relic.

You can access New Relic through your browser, and we will learn how to make a New Relic dashboard by taking the following steps:

1. In a browser, open the New Relic website: `https://infrastructure.newrelic.com/`.

2. Next, you need to sign in. If you do not have an account, you can register with New Relic by clicking on **Create a free account**:

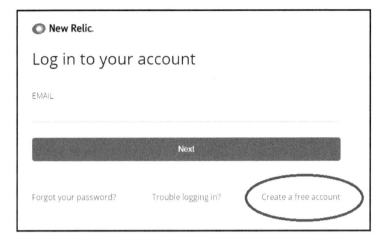

Figure 9-1. New Relic login

3. Enter the required information into the New Relic registration form as shown in *Figure 9-2* and *Figure 9-3*:

Figure 9-2. New Relic user registration

Scroll down to the end of the web page to check the **Terms of Service** and **Service Policy Notice** before you click the **Start now** button:

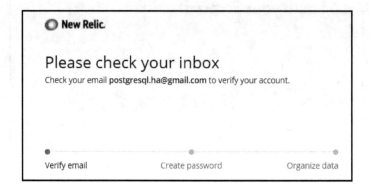

Figure 9-3. New Relic user registration (cont.)

4. You should receive a notification about verification as shown here:

Figure 9-4. New Relic registration email

5. When you click on the link inside the New Relic email, it will pop up the **Create your password** screen for you, as shown in Figure 9-5:

Figure 9-5. Setting a New Relic password

For example, I set up my password as BookDemo@9.

6. Now with your email ID and password, you need to sign in here:

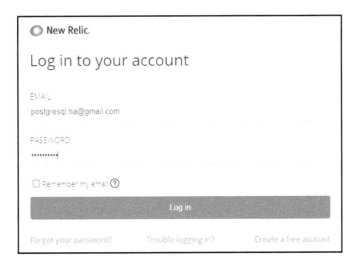

Figure 9-6. Signing in to New Relic as a newly registered user

7. After you click on the **Log in** button, the New Relic APM page will open. Select the Data Region as either **United States** or **Europe**:

Figure 9-7. New Relic APM page

8. Click on the **Save** button:

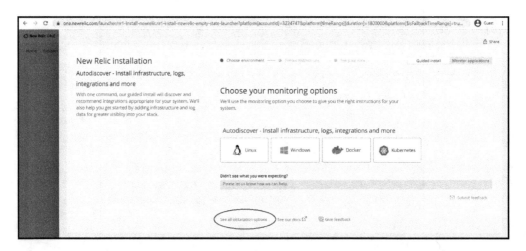

Figure 9-8. New Relic installation page

9. Click on the **See all installation options** link in the previous figure and select the infrastructure type as Amazon Web Services as seen here:

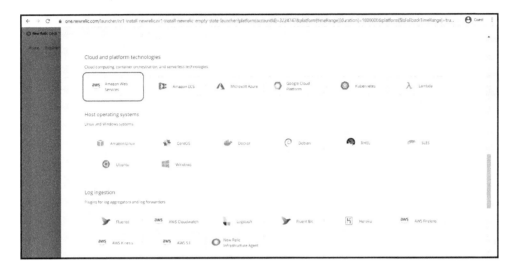

Figure 9-9. Different New Relic infrastructure types

10. Select the type of AWS services that you would like to set up. Here, I am going to select **RDS** for the PostgreSQL database:

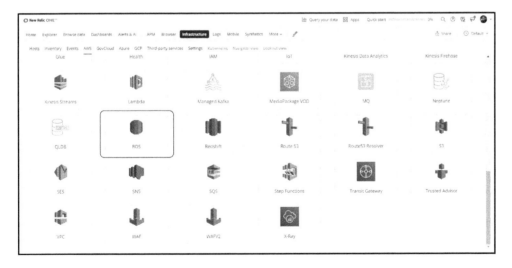

Figure 9-10. RDS inside New Relic infrastructure

Not only can New Relic offer a monitoring dashboard for PostgreSQL RDS but it can also visualize monitoring for many other AWS services, such as IoT, Kinesis, Lambda, Redshift, and VPC. By simply signing up for a new user account, you can start to monitor different AWS services with a few easy steps.

11. There are 2 integration modes for AWS, please press the **Use API Polling** button.

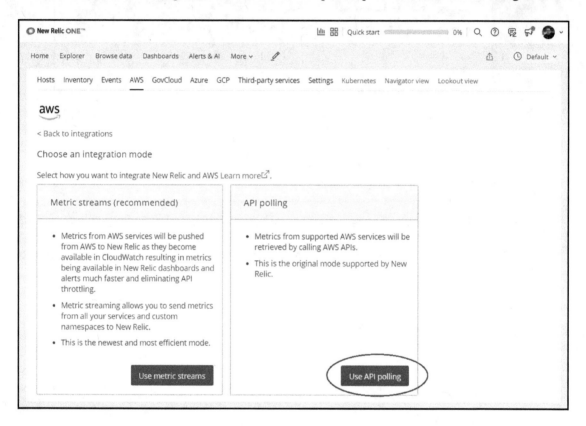

Fig 9.11 Integration modes for RDS

In order to connect your New Relic account to the PostgreSQL RDS, you have to grant permissions from your AWS account so that New Relic will be allowed to collect information about your RDS.

We will discuss the required role permissions that we need to set up for New Relic to access your RDS in the next section.

Defining PostgreSQL role permissions

AWS uses the **Identity and Access Management** (**IAM**) service to enable you to grant access to AWS services and resources. With IAM, you can create a new role using permissions to allow New Relic access to AWS resources. Therefore, you need to open two tabs at the same time on your browser for both AWS and New Relic, then you have to copy values from your New Relic account into the AWS IAM service.

The following are the steps to create an IAM policy:

1. Now we will link the New Relic user account to the PostgreSQL RDS. In Step 1 on the New Relic page called Trust, remember to copy the details on the page and open a new browser window/tab for AWS. You will then switch between the two browser windows – the browser window of New Relic and the browser window of AWS:

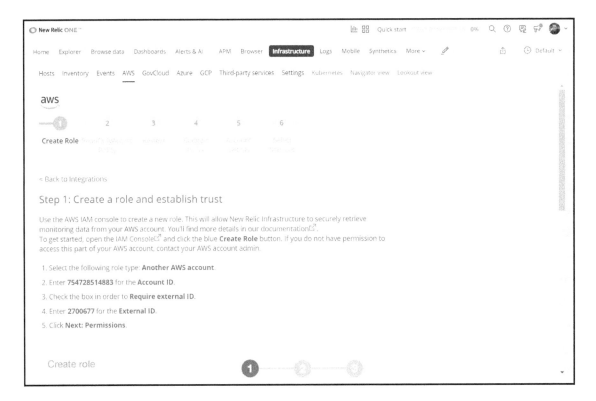

Figure 9-12. The first step – trust

2. In the second browser window, please sign in and navigate to the AWS Management Console using this link: `https://console.aws.amazon.com/iam/home#roles`. Click on **Create role** to set up an IAM role for granting permissions to New Relic to access our AWS services:

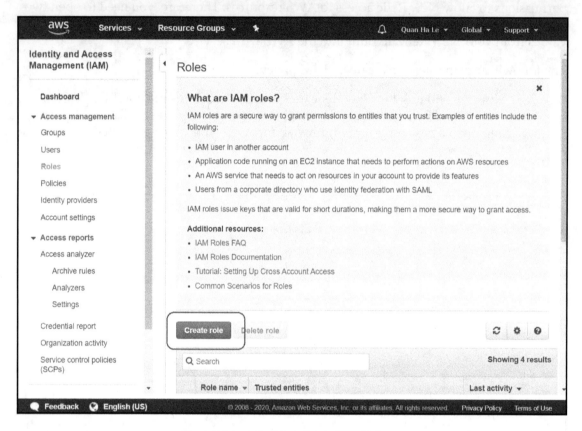

Figure 9-13. Repeat the New Relic details on the AWS IAM role page

3. Switch back to the first New Relic window and scroll down the page to see all the guidelines for step 1:

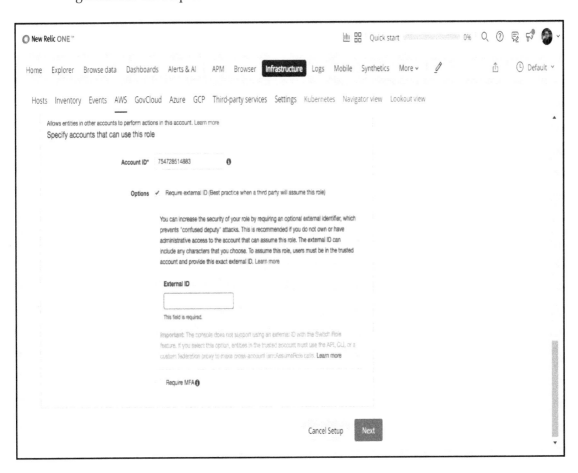

Figure 9-14. New Relic step 1 – trust (cont.)

4. Switch to the AWS browser to complete step 2. We'll now select the type of trusted entity as **Another AWS account** in the AWS browser window. Then copy the values from the first browser window of New Relic for the following values on the AWS page and then click on **Next: Permissions**:

- **Account ID**: 754728514883
- **Require external ID**: Yes
- **External ID**: 2700677

We can see this in the following screenshot:

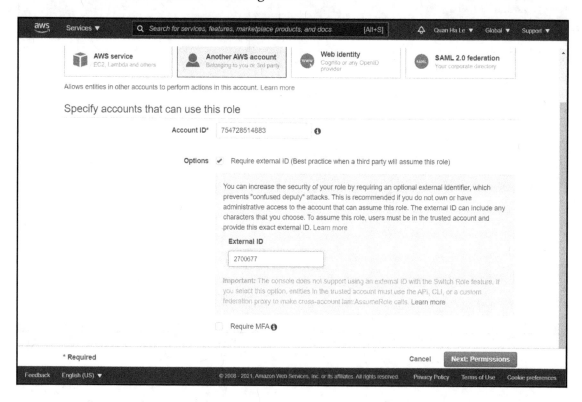

Figure 9-15. Type of trusted entity

5. In the New Relic window, click **Next** to move on to the next guideline step, **Permissions**:

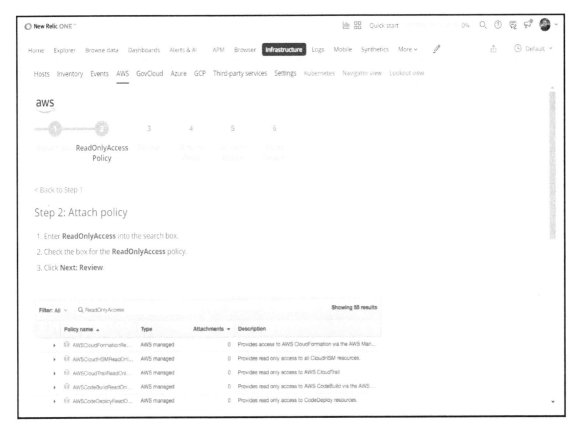

Figure 9-16. New Relic step 2 – Permissions

6. We will switch to the AWS window and will follow the New Relic
    guidelines from step 5 to proceed further:

    1. Enter `ReadOnlyAccess` into the search box.

    2. Scroll down until nearly the end of the list, then check the
       box for **ReadOnlyAccess**.

    3. Click the **Next: Tags** button:

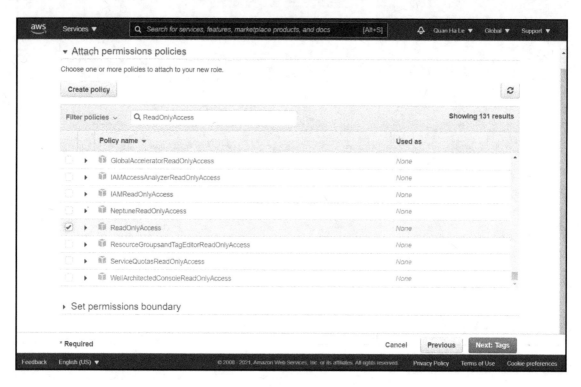

Figure 9-17. AWS permissions

7. Click the **Next: Review** button:

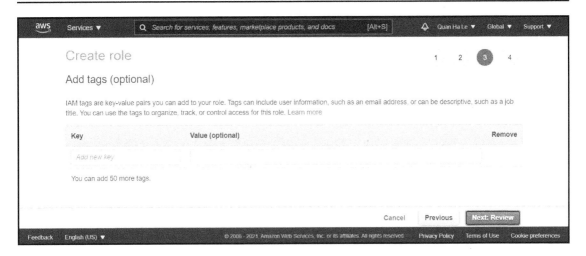

Figure 9-18. No tags are needed

8. In the New Relic window, click **Next** to move on to the next guidelines, step 3, called **Review**:

Figure 9-19. New Relic step 3 – Review

9. In the AWS window, set a name for the IAM role and click on **Create role**:

Figure 9-20. Integration of the AWS role and New Relic

10. Click on the new role name:

Figure 9-21. A New IAM role for New Relic has been created

11. Copy the IAM role ARN by clicking on the 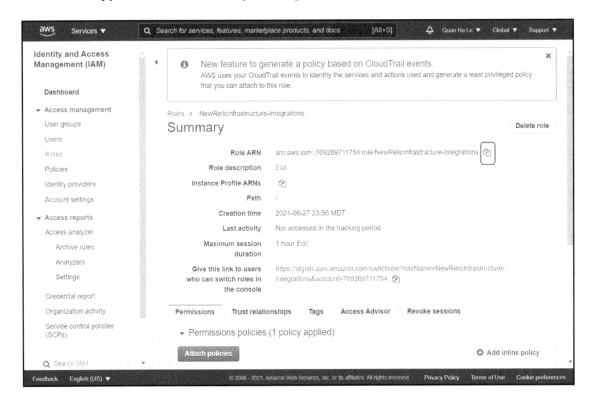 icon:

Figure 9-22. The New Relic role ARN

Keep the role ARN somewhere safe to use later.

12. Now switch to the New Relic window and click on **Next** to move on to step 4, which is **Budgets Policy**:

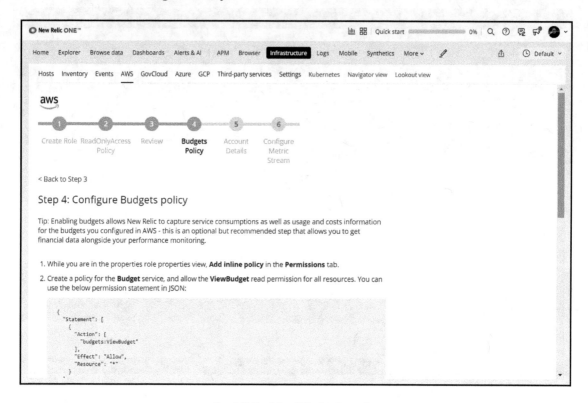

Figure 9-23. New Relic guideline for policy creation

13. In the AWS window, click on **Add inline policy** (Figure 9-22) and then click on **Choose a service**:

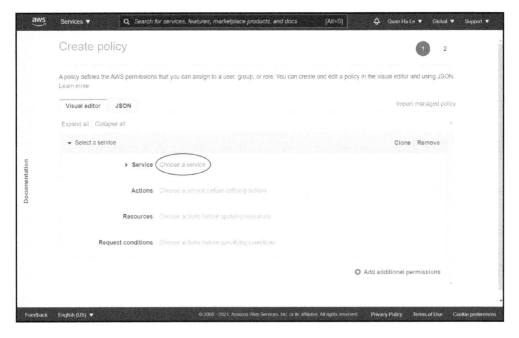

Figure 9-24. Create policy page

14. Enter `Budget` to select the **Budget** service:

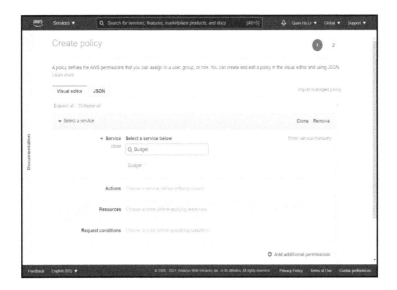

Figure 9-25. Select the Budget service

15. The **Actions** area will expand. For **Access level**, select **Read**:

Figure 9-26. Policy access level

Then, for **Resources**, select **All resources** and click on the **Review policy** button:

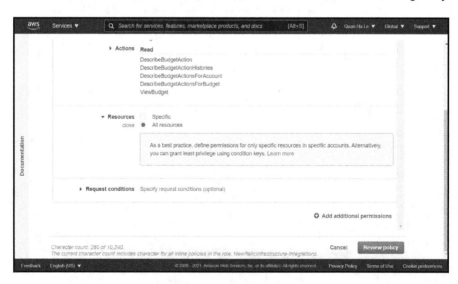

Figure 9-27. Budget resources

16. Enter the policy name in the **Name** field as **NewRelicBudget** and then click on the **Create policy** button:

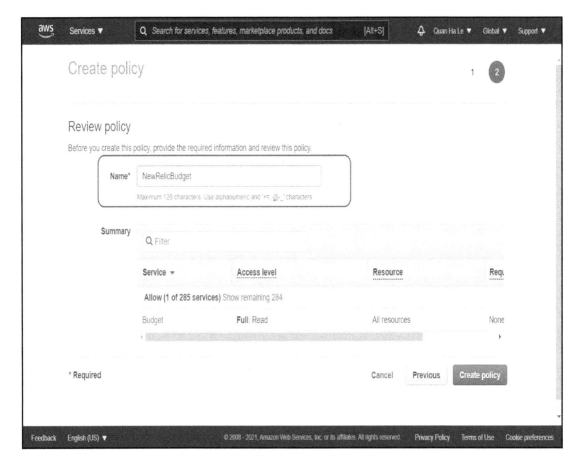

Figure 9-28. Create policy

In this section, you learned how to grant permissions for New Relic to access the AWS RDS service. The steps are straightforward using IAM roles of AWS. New Relic also provides step-by-step detailed guidance, which is very convenient for system administrators.

Now that you have prepared your AWS account, in the next section, you will learn how to link it with your New Relic account to use the created IAM **NewRelicBudget** policy.

Configuring New Relic for PostgreSQL

Now it is time to provide your AWS account details for New Relic. Once New Relic connects to your AWS RDS service using these details, because you have already set up IAM role permissions to authorize New Relic, the New Relic dashboard will be able to launch:

1. Switch to the New Relic window and click on **Next** to move on to the next step, **Account Details,** in the New Relic browser:

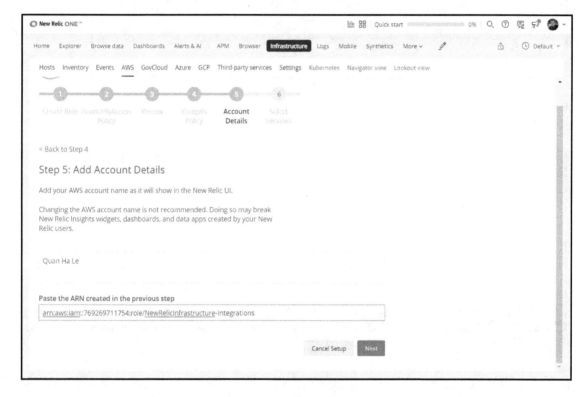

Figure 9-29. Step 5 Account Details

In Figure 9-30, we added the AWS account name in New Relic, and then for the IAM Role ARN, we have to find the correct values from the AWS window.

2. Switch to the AWS window and click on the **My Account** drop-down field to get the AWS account details for New Relic as shown in Figure 9-30:

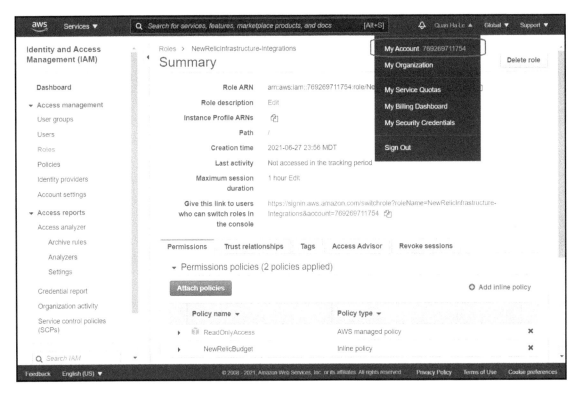

Figure 9-30. AWS account page

3. Copy the account name as shown in Figure 9-31:

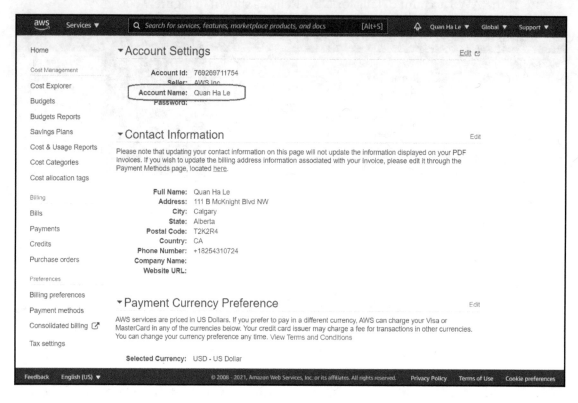

Figure 9-31. AWS My Account page

4. In the New Relic window, click **Next** to proceed to step 6, which is **Select Services**:

Figure 9-32. Step 6 – Select Services

5. Next, you need to de-select the **Select all** checkbox and then select the **RDS** option:

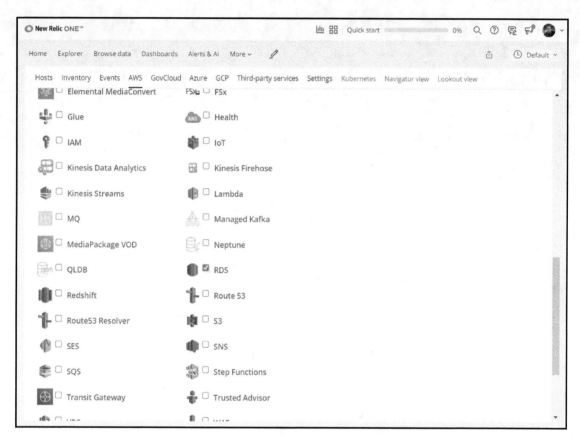

Figure 9-33. New Relic RDS option

6. Click on **Next** and then we will reach our final results:

Figure 9-34. Your AWS services with a New Relic infrastructure

7. Click on **RDS dashboard** (Figure 9-34). After around 10 to 15 minutes, you will observe that the monitoring data will be captured from AWS PostgreSQL RDS in New Relic as shown in Figure 9-35:

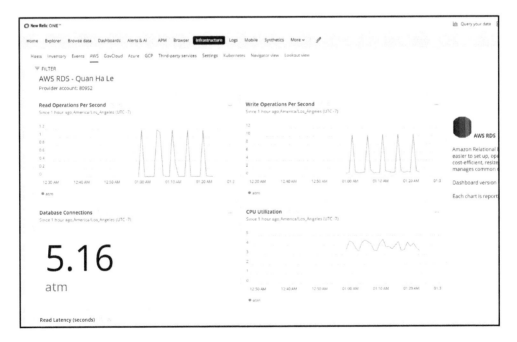

Figure 9-35. RDS dashboard of New Relic

New Relic uses an API as an intermediate layer between AWS RDS and the New Relic dashboard so that the charts in Figure 9-35 can be shown. The data sent from this API is called *metric data* hence the API can be called the *Metric API*. In the next section, we will go through some examples of New Relic metric data.

Adding new metric data for PostgreSQL

The Metric API is a way to get metric data from AWS PostgreSQL RDS into New Relic dashboards. A metric means a numeric measurement of an application, a system, or a database. Metrics are typically reported on a regular schedule. There are two types of metrics:

- Aggregated data: the rate of some event per minute
- A numeric status at a moment in time: for example, the CPU% used status

Metrics are a strong solution for monitoring dashboards. The metrics that are used in New Relic are as follows:

- readIops: This measures the average number of disk I/O operations per second.
- writeIops: This measures the average number of disk I/O operations per second.
- databaseConnections: This measures the number of connections to an instance.
- cpuUtilization: This measures the percentage of CPU used by a DB instance.
- readLatency: This measures the average amount of time taken per disk I/O operation, in seconds.
- writeLatency: This measures the average amount of time taken per disk I/O operation.
- networkReceiveThroughput: This measures the amount of network throughput received from clients by each instance in the Aurora MySQL DB cluster, in bytes per second.
- networkTransmitThroughput: This measures the amount of network throughput sent to clients by each instance in the Aurora DB cluster, in bytes per second.
- swapUsageBytes: This measures the amount of swap space used on the Aurora PostgreSQL DB instance, in bytes.

We will show you an example of `readIops` and `readLatency` in this section:

1. In order to see how New Relic uses the `readIops` metric, click on the **...** button on the **Read Operations Per Second** chart, then select **Copy to query builder**:

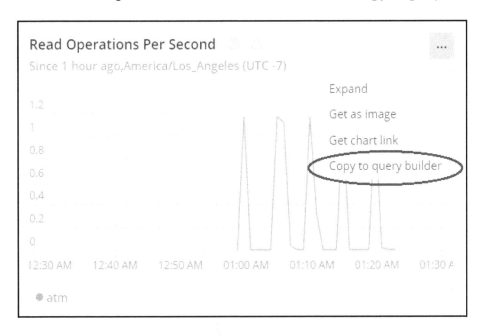

Figure 9-36. Read Operations Per Second chart

2. You will see the output of the following `SELECT` query in Figure 9-36. This query selects the average value of `readIops` measurements of our PostgreSQL RDS through the Metric API:

```
SELECT average(`provider.readIops.Average`) as 'Read Operations'
From DatastoreSample WHERE provider = 'RdsDbInstance' AND
providerAccountId = '34337' Since 355 minutes ago TIMESERIES Until
10 minutes ago facet displayName
```

Hence the values are an average of the read input/output operations per second:

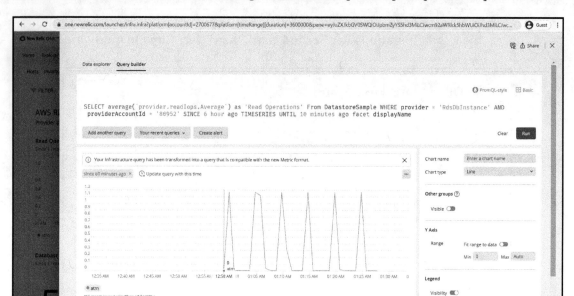

Figure 9-37. Average of readIops

3. Another example for the usage of the `readLatency` metric is with the `sum()` function. We will take the same steps but we will change the query:

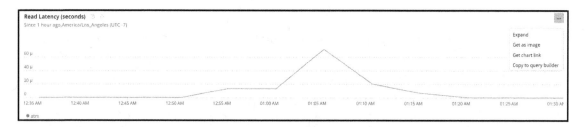

Figure 9-38. Read Latency chart

Here is the New Relic query that uses the `sum()` function:

```
SELECT sum(`provider.readLatency.Sum`) / 60 as 'seconds' From
DatastoreSample WHERE provider = 'RdsDbInstance' AND
providerAccountId = '34337' timeseries 5 minutes Until 10 minutes
ago Since 6 hours ago facet displayName
```

As for the previous query, we will copy it inside a modal window:

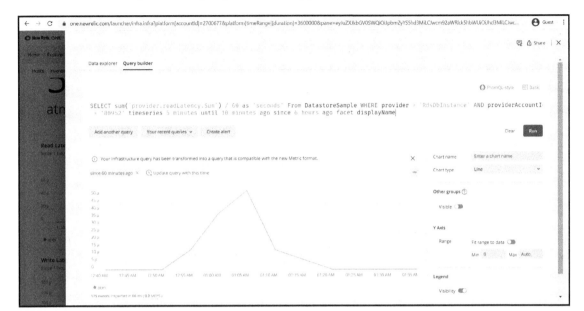

Figure 9-39. Sum of readLatency

In this section, you were introduced to the mechanism of how New Relic measures activities from your PostgreSQL RDS, which is metric data. You can open any charts from your New Relic dashboard of PostgreSQL RDS and try to view its query to get more experience with the usage of metric data.

Infrastructure inventory data collection

Each AWS service also has its own parameters and values defined by AWS settings. New Relic calls the AWS parameters as an *infrastructure inventory data collection*. So New Relic will also show the resource parameters on the dashboards so that the system administrator can easily look for the AWS settings and values directly in New Relic.

The **RDS Inventory** area is located at the end of the RDS dashboard of New Relic, as shown in Figure 9-40:

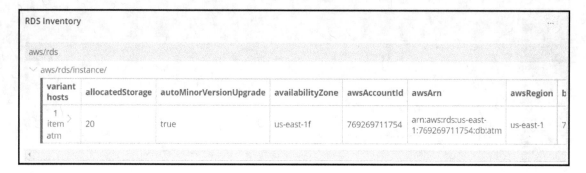

Figure 9-40. RDS Inventory

You can scroll right and left to see all of the inventory properties of the RDS. All of the property names and explanations are in the following table:

Name	Description
awsRegion	The AWS region that the instance was provisioned in.
dbInstanceIdentifier	Contains a user-supplied database identifier. This identifier is the unique key that identifies a DB instance.
allocatedStorage	Specifies the allocated storage size, in gibibytes.
autoMinorVersionUpgrade	Indicates that minor version patches are applied automatically.
availabilityZone	Specifies the name of the Availability Zone the DB instance is located in.
backupRetentionPeriod	Specifies the number of days for which automatic DB snapshots are retained.
caCertificateIdentifier	The identifier of the CA certificate for this DB instance.
characterSetName	If present, specifies the name of the character set that this instance is associated with.
dbClusterIdentifier	If the DB instance is a member of a DB cluster, contains the name of the DB cluster that the DB instance is a member of.
dbInstanceClass	Contains the name of the compute and memory capacity class of the DB instance.

Name	Description
dbInstancePort	Specifies the port that the DB instance listens on. If the DB instance is part of a DB cluster, this can be a different port than the DB cluster port.
dbInstanceStatus	Specifies the current state of this database.
dbName	Contains the name of the initial database of this instance that was provided at creation time, if one was specified when the DB instance was created. This same name is returned for the life of the DB instance.
dbParameterGroups	Provides the list of DB parameter groups applied to this DB instance.
dbSecurityGroups	Provides a list of DB security group elements.
dbSubnetGroup	Specifies information on the subnet group associated with the DB instance, including the name, description, and subnets in the subnet group.
endpoint	Specifies the connection endpoint.
engine	Provides the name of the database engine to be used for this DB instance.
engineVersion	Indicates the database engine version.
kmsKeyId	If StorageEncrypted is true, the AWS KMS key identifier for the encrypted DB instance.
licenseModel	License model information for this DB instance.
masterUsername	Contains the master username for the DB instance.
multiAz	Specifies whether the DB instance is a Multi-AZ deployment.
optionGroupMemberships	Provides the list of option group memberships for this DB instance.
preferredBackupWindow	Specifies the daily time range during which automated backups are created if automated backups are enabled, as determined by BackupRetentionPeriod.
preferredMaintenanceWindow	Specifies the weekly time range during which system maintenance can occur, in **Universal Coordinated Time (UTC)**.
publiclyAccessible	Specifies the accessibility options for the DB instance. A value of true specifies an internet-facing instance with a publicly resolvable DNS name, which resolves to a public IP address. A value of false specifies an internal instance with a DNS name that resolves to a private IP address.

Name	Description
readReplicaDdInstanceIdentifiers	Contains one or more identifiers of the read replicas associated with this DB instance.
readReplicaSourceDbInstanceIdentifier	Contains the identifier of the source DB instance if this DB instance is a read replica.
secondaryAvailabilityZone	If present, specifies the name of the secondary Availability Zone for a DB instance with multi-AZ support.
storageEncrypted	Specifies whether the DB instance is encrypted.
storageType	Specifies the storage type associated with the DB instance.
tdeCredentialArn	The ARN from the key store with which the instance is associated for TDE encryption.
vpcSecurityGroups	Provides a list of VPC security group elements that the DB instance belongs to.
clusterInstance	Specifices whether the instance is a cluster or not.
tags*	Instance tags.

Table 9-1. RDS Inventory properties

The inventory area of New Relic provides a real-time, filterable, searchable view into the PostgreSQL RDS parameters. If you change any settings on your PostgreSQL RDS by using the AWS service, you will immediately see the new values updated on the New Relic inventory. New Relic is very convenient and easy to use, so whenever you need to, you can look directly at the New Relic dashboard instead of opening the AWS service to search for a database parameter.

Summary

In this chapter, we went through step-by-step PostgreSQL RDS database monitoring through New Relic, which is a popular DevOps tool. We learned how to register the Amazon RDS monitoring integration of New Relic. Going ahead, we learned how to prepare the IAM role for New Relic to collect PostgreSQL metrics. We also investigated the RDS dashboard of New Relic. Finally, we understood the metric data for PostgreSQL and learned how to read the RDS inventory data.

This chapter focused on PostgreSQL monitoring for high-performance teams. In the next chapter, we will move on to PostgreSQL database performance monitoring with PGBench and Apache JMeter.

Testing the Performance of Our Banking App with PGBench and JMeter

In this chapter, we will learn how to create a load test for a PostgreSQL database to benchmark PostgreSQL performance with pgbench and JMeter. The purpose of benchmarking a database is not only to check the capability of the database but also to check the behavior of a particular database against your application. Different hardware provides different results based on the benchmarking plan that you set. It is very important to isolate the server (the actual one being benchmarked) from other elements such as the servers driving the load or the servers used to collect and store performance metrics. After benchmarking, you will also get application characteristics.

With the help of a project in this chapter, we will create and run load tests to benchmark the performance of the PostgreSQL 12 RDS from Amazon Web Services from Chapter 2, *Creating a Geospatial Database Using PostGIS and PostgreSQL*, related to bank ATM locations in a typical city.

The following topics will be covered in this chapter:

- How to benchmark PostgreSQL performance
- pgbench 1 – Creating and initializing a benchmark database
- pgbench 2 – Running a baseline pgbench test
- pgbench 3 – Creating and testing a connection pool
- JMeter setup
- JMeter for AWS PostgreSQL RDS

Technical requirements

This chapter will take developers around 10-12 hours of work to develop PostgreSQL performance tests.

How to benchmark PostgreSQL performance

We usually have to benchmark our PostgreSQL database to make sure that before we deliver our products for customers, the capability of the PostgreSQL database will be able to cover the expected traffic loads in production.

Benchmarks are super important in our initial implementation stage since we can estimate and answer questions such as the following:

- How many users can we support?
- What response time will we have at peak usage time?
- Will the storage space have enough capacity?
- Will the current hardware meet expectations?

Based on the answers that we obtain, we can adjust tuning parameters in PostgreSQL and at the same time, estimate what hardware we will need in the future.

Having said all that, one of the common ways to benchmark PostgreSQL is through two typical tools:

- pgbench, a built-in benchmarking tool of PostgreSQL
- Apache JMeter software, a tool designed to load test functional behavior and measure performance

In the following topics, we will perform some strength tests on our ATM database on PostgreSQL RDS and we will see its behavior through those benchmarking tools.

pgbench 1 – Creating and initializing a benchmark database

We are going to use pgbench for the RDS on Amazon Web Services. We will set up pgbench from the Jenkins server at `192.168.0.200` that we set up with Vagrant in Chapter 7, *PostgreSQL with DevOps for Continuous Delivery*. We PuTTY into the Jenkins server:

1. First, open PowerShell as administrator:

```
PS C:\Windows\system32>
PS C:\Windows\system32> cd C:\Projects\Vagrant\Jenkins
PS C:\Projects\Vagrant\Jenkins> vagrant up --provider virtualbox
PS C:\Projects\Vagrant\Jenkins> vagrant ssh

vagrant@devopsubuntu1804:~$
```

2. Execute the following script to install pgbench:

```
vagrant@devopsubuntu1804:~$ wget --quiet -O -
https://www.postgresql.org/media/keys/ACCC4CF8.asc | sudo apt-key
add -

vagrant@devopsubuntu1804:~$ echo "deb
http://apt.postgresql.org/pub/repos/apt/ `lsb_release -cs`-pgdg
main" |sudo tee /etc/apt/sources.list.d/pgdg.list

vagrant@devopsubuntu1804:~$ sudo apt update
vagrant@devopsubuntu1804:~$ sudo apt -y install postgresql-contrib
vagrant@devopsubuntu1804:~$ pgbench -V
pgbench (PostgreSQL) 12.2 (Ubuntu 12.2-2.pgdg18.04+1)
```

3. Now we will initialize the benchmark database by using the following initialization syntax:

```
Syntax:
pgbench -i [-h hostname] [-p port] [-s scaling_factor] [-U login]
[-d] [dbname]
```

The parameter definitions are given as follows:

- `scaling_factor`: The number of tuples generated will be a multiple of the scaling factor. For example, `-s 1` will imply 100,000 tuples in the `pgbench_accounts` table.
- `port`: Default 5432.
- `hostname`: Default localhost.
- `-d`: Shows debug information.

4. Hence we issue the `pgbench` command to our RDS endpoint with 1 million tuples:

```
vagrant@devopsubuntu1804:~$ pgbench -i -h atm.ck5074bwbilj.us-
east-1.rds.amazonaws.com -p 5432 -s 10 -U dba -d atm
Password: (please enter the RDS password = bookdemo)
dropping old tables...
NOTICE: table "pgbench_accounts" does not exist, skipping
NOTICE: table "pgbench_branches" does not exist, skipping
NOTICE: table "pgbench_history" does not exist, skipping
NOTICE: table "pgbench_tellers" does not exist, skipping
creating tables...
generating data...
100000 of 1000000 tuples (10%) done (elapsed 1.93 s, remaining
17.34 s)
200000 of 1000000 tuples (20%) done (elapsed 4.27 s, remaining
17.09 s)
300000 of 1000000 tuples (30%) done (elapsed 6.82 s, remaining
15.91 s)
400000 of 1000000 tuples (40%) done (elapsed 9.46 s, remaining
14.19 s)
500000 of 1000000 tuples (50%) done (elapsed 12.12 s, remaining
12.12 s)
600000 of 1000000 tuples (60%) done (elapsed 14.49 s, remaining
9.66 s)
700000 of 1000000 tuples (70%) done (elapsed 17.13 s, remaining
7.34 s)
800000 of 1000000 tuples (80%) done (elapsed 19.80 s, remaining
4.95 s)
900000 of 1000000 tuples (90%) done (elapsed 22.16 s, remaining
2.46 s)
1000000 of 1000000 tuples (100%) done (elapsed 24.83 s, remaining
0.00 s)
vacuuming...
creating primary keys...
done.
```

Please look at *Figure 10.1* for an illustration:

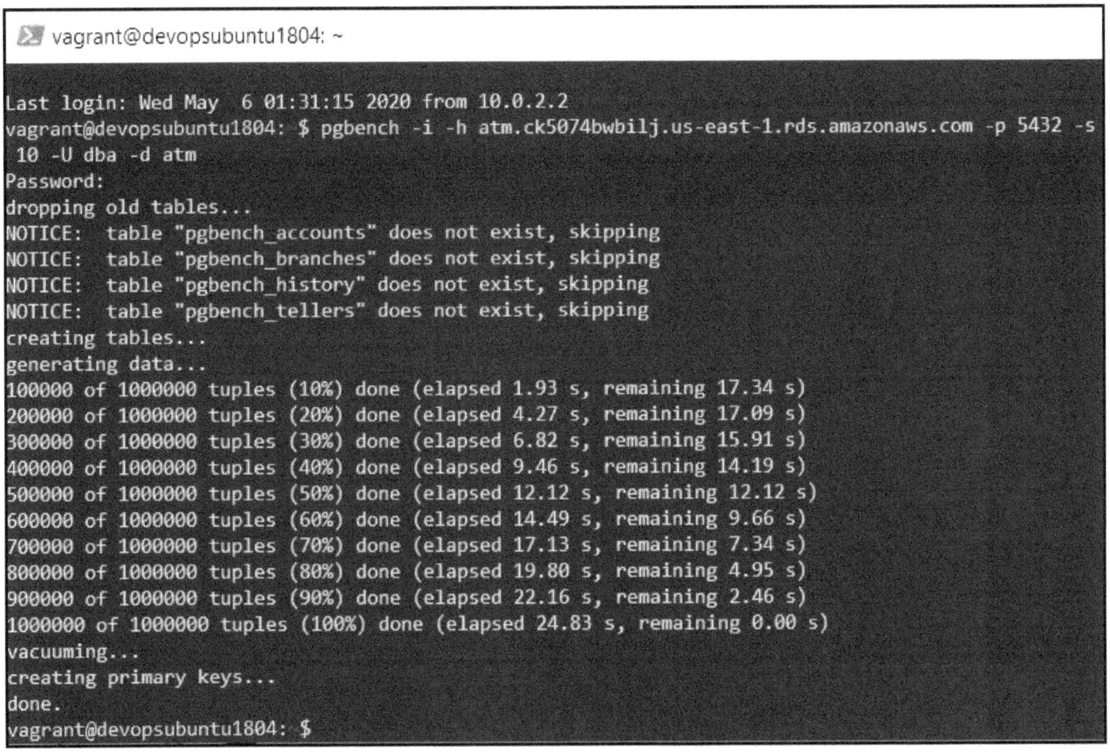

```
vagrant@devopsubuntu1804: ~

Last login: Wed May  6 01:31:15 2020 from 10.0.2.2
vagrant@devopsubuntu1804: $ pgbench -i -h atm.ck5074bwbilj.us-east-1.rds.amazonaws.com -p 5432 -s
 10 -U dba -d atm
Password:
dropping old tables...
NOTICE:  table "pgbench_accounts" does not exist, skipping
NOTICE:  table "pgbench_branches" does not exist, skipping
NOTICE:  table "pgbench_history" does not exist, skipping
NOTICE:  table "pgbench_tellers" does not exist, skipping
creating tables...
generating data...
100000 of 1000000 tuples (10%) done (elapsed 1.93 s, remaining 17.34 s)
200000 of 1000000 tuples (20%) done (elapsed 4.27 s, remaining 17.09 s)
300000 of 1000000 tuples (30%) done (elapsed 6.82 s, remaining 15.91 s)
400000 of 1000000 tuples (40%) done (elapsed 9.46 s, remaining 14.19 s)
500000 of 1000000 tuples (50%) done (elapsed 12.12 s, remaining 12.12 s)
600000 of 1000000 tuples (60%) done (elapsed 14.49 s, remaining 9.66 s)
700000 of 1000000 tuples (70%) done (elapsed 17.13 s, remaining 7.34 s)
800000 of 1000000 tuples (80%) done (elapsed 19.80 s, remaining 4.95 s)
900000 of 1000000 tuples (90%) done (elapsed 22.16 s, remaining 2.46 s)
1000000 of 1000000 tuples (100%) done (elapsed 24.83 s, remaining 0.00 s)
vacuuming...
creating primary keys...
done.
vagrant@devopsubuntu1804: $
```

Figure 10.1. pgbench initialization

After the `pgbench` command has been completed with the parameters that we defined previously, we can open the RDS database by using pgAdmin and we will see that there are four more pgbench tables added – **pgbench_accounts**, **pgbench_branches**, **pgbench_history**, and **pgbench_tellers**:

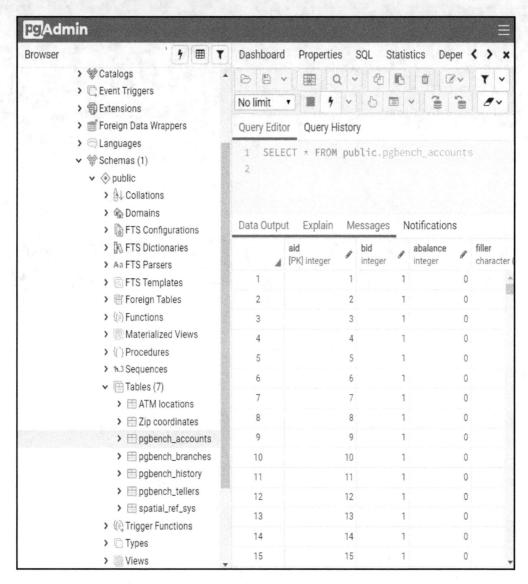

Figure 10.2. pgbench tables

Obviously, from pgAdmin, we can easily check that there are 1 million rows generated in the table `pgbench_accounts`.

pgbench 2 – Running a baseline pgbench test

1. This is the general syntax to run a pgbench benchmarking test:

```
Syntax:
pgbench  [-h hostname]  [-p port]  [-c nclients]  [-t
ntransactions]  [-s scaling_factor] [-D varname=value]  [-n]  [-C]
[-v] [-S]  [-N]  [-f filename]  [-j threads]  [-l]  [-U login]  [-
d]  [dbname]
```

We'll see the following parameters:

- `nclients`: Client number – default 1
- `ntransactions`: Transactions per client – default 10
- `varname=value`: Defined variable referred to by the file script
- `-n`: novacuuming – cleaning the history table before testing
- `-C`: One connection for each transaction
- `-v`: Vacuuming before testing
- `-S`: Only SELECT transactions
- `-N`: No updates on the `pgbench_branches` and `pgbench_tellers` tables
- `filename`: Read the transaction script from a specified SQL file
- `threads`: The number of worker threads
- `-l`: Write the time of each transaction to a `pgbench_log` file

2. Hence we can create a small load of 50 users. Each user runs 100 transactions with 2 threads on the RDS `atm` database:

```
vagrant@devopsubuntu1804:~$ pgbench -h atm.ck5074bwbilj.us-
east-1.rds.amazonaws.com -U dba -c 50 -j 2 -t 100 -d atm
pghost: atm.ck5074bwbilj.us-east-1.rds.amazonaws.com pgport:
nclients: 50 nxacts: 100 dbName: atm
Password: (please enter the RDS password = bookdemo)
...
...
client 49 executing script "<builtin: TPC-B (sort of)>"
client 49 executing \set aid
client 49 executing \set bid
client 49 executing \set tid
```

```
client 49 executing \set delta
client 49 sending BEGIN;
client 49 receiving
client 49 receiving
client 49 sending UPDATE pgbench_accounts SET abalance = abalance +
-1345 WHERE aid = 938376;
client 49 receiving
client 49 receiving
client 49 sending SELECT abalance FROM pgbench_accounts WHERE aid =
938376;
client 49 receiving
client 49 receiving
client 49 sending UPDATE pgbench_tellers SET tbalance = tbalance +
-1345 WHERE tid = 63;
client 49 receiving
client 49 receiving
client 49 sending UPDATE pgbench_branches SET bbalance = bbalance +
-1345 WHERE bid = 1;
client 49 receiving
client 49 receiving
client 49 sending INSERT INTO pgbench_history (tid, bid, aid,
delta, mtime) VALUES (63, 1, 938376, -1345, CURRENT_TIMESTAMP);
client 49 receiving
client 49 receiving
client 49 sending END;
client 49 receiving
client 49 receiving
transaction type: <builtin: TPC-B (sort of)>
scaling factor: 10
query mode: simple
number of clients: 50
number of threads: 2
number of transactions per client: 100
number of transactions actually processed: 5000/5000
latency average = 1008.361 ms
tps = 49.585429 (including connections establishing)
tps = 49.797092 (excluding connections establishing)
```

Therefore, the total number of transactions will be 50 users x 100 transactions = 5,000 transactions. The throughput of pgbench shows that less than 50 transactions are completed per second:

```
 vagrant@devopsubuntu1804: ~
client 10 sending END;
client 10 receiving
client 49 receiving
client 49 sending END;
client 49 receiving
client 10 receiving
client 49 receiving
client 49 executing script "<builtin: TPC-B (sort of)>"
client 49 executing \set aid
client 49 executing \set bid
client 49 executing \set tid
client 49 executing \set delta
client 49 sending BEGIN;
client 49 receiving
client 49 receiving
client 49 sending UPDATE pgbench_accounts SET abalance = abalance + -1345 WHERE aid = 938376;
client 49 receiving
client 49 receiving
client 49 sending SELECT abalance FROM pgbench_accounts WHERE aid = 938376;
client 49 receiving
client 49 receiving
client 49 sending UPDATE pgbench_tellers SET tbalance = tbalance + -1345 WHERE tid = 63;
client 49 receiving
client 49 receiving
client 49 sending UPDATE pgbench_branches SET bbalance = bbalance + -1345 WHERE bid = 1;
client 49 receiving
client 49 receiving
client 49 sending INSERT INTO pgbench_history (tid, bid, aid, delta, mtime) VALUES (63, 1, 938376
, -1345, CURRENT_TIMESTAMP);
client 49 receiving
client 49 receiving
client 49 sending END;
client 49 receiving
client 49 receiving
transaction type: <builtin: TPC-B (sort of)>
scaling factor: 10
query mode: simple
number of clients: 50
number of threads: 2
number of transactions per client: 100
number of transactions actually processed: 5000/5000
latency average = 1008.361 ms
tps = 49.585429 (including connections establishing)
tps = 49.797092 (excluding connections establishing)
vagrant@devopsubuntu1804: $ _
```

Figure 10.3. pgbench performance for 50 users

3. Now let's run the same test. This time, we will use a number of users higher than our RDS allowance. Hence, from the pgAdmin browser, please select the atm database and execute the following `SELECT` statement:

```
select * from pg_settings where name='max_connections';
```

The value of `max_connections` of the PostgreSQL RDS is `87` for the atm database, as shown in the following screenshot. That capacity is correct for our RDS size – `db.t2.micro`:

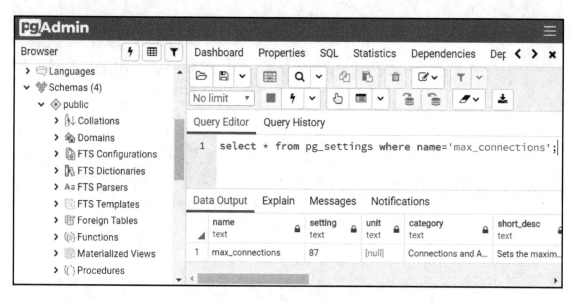

Figure 10.4. Connection allowance of the db.t2.micro PostgreSQL RDS

4. We will now use 150 concurrent users for our next pgbench test – a value that is higher than `max_connections=87` for this PostgreSQL database – to synthetically simulate a mass influx of client connections:

```
vagrant@devopsubuntu1804:~$ pgbench -h atm.ck5074bwbilj.us-
east-1.rds.amazonaws.com -U dba -c 150 -j 2 -t 100 -d atm
pghost: atm.ck5074bwbilj.us-east-1.rds.amazonaws.com pgport:
nclients: 150 nxacts: 100 dbName: atm
Password: (please enter the RDS password = bookdemo)
starting vacuum...end.
connection to database "atm" failed:
FATAL: remaining connection slots are reserved for non-replication
superuser connections
...
...
```

```
client 37 executing script "<builtin: TPC-B (sort of)>"
client 37 executing \set aid
client 37 executing \set bid
client 37 executing \set tid
client 37 executing \set delta
client 37 sending BEGIN;
client 37 receiving
client 37 receiving
client 37 sending UPDATE pgbench_accounts SET abalance = abalance +
4605 WHERE aid = 560242;
client 37 receiving
client 37 receiving
client 37 sending SELECT abalance FROM pgbench_accounts WHERE aid =
560242;
client 37 receiving
client 37 receiving
client 37 sending UPDATE pgbench_tellers SET tbalance = tbalance +
4605 WHERE tid = 20;
client 37 receiving
client 37 receiving
client 37 sending UPDATE pgbench_branches SET bbalance = bbalance +
4605 WHERE bid = 7;
client 37 receiving
client 37 receiving
client 37 sending INSERT INTO pgbench_history (tid, bid, aid,
delta, mtime) VALUES (20, 7, 560242, 4605, CURRENT_TIMESTAMP);
client 37 receiving
client 37 receiving
client 37 sending END;
client 37 receiving
client 37 receiving
transaction type: <builtin: TPC-B (sort of)>
scaling factor: 10
query mode: simple
number of clients: 150
number of threads: 2
number of transactions per client: 100
number of transactions actually processed: 7500/15000
latency average = 3138.897 ms
tps = 47.787488 (including connections establishing)
tps = 47.853332 (excluding connections establishing)
```

Note the **FATAL** error, indicating that pgbench hits the 87-connection limit threshold set by `max_connections`, resulting in a refused connection. The test was still able to complete, with a throughput of less than 48 as follows:

```
vagrant@devopsubuntu1804: ~                                      —    □    ×
, -84, CURRENT_TIMESTAMP);
client 37 receiving
client 66 receiving
client 37 receiving
client 37 sending END;
client 37 receiving
client 37 receiving
client 37 executing script "<builtin: TPC-B (sort of)>"
client 37 executing \set aid
client 37 executing \set bid
client 37 executing \set tid
client 37 executing \set delta
client 37 sending BEGIN;
client 37 receiving
client 37 receiving
client 37 sending UPDATE pgbench_accounts SET abalance = abalance + 4605 WHERE aid = 560242;
client 37 receiving
client 37 receiving
client 37 sending SELECT abalance FROM pgbench_accounts WHERE aid = 560242;
client 37 receiving
client 37 receiving
client 37 sending UPDATE pgbench_tellers SET tbalance = tbalance + 4605 WHERE tid = 20;
client 37 receiving
client 37 receiving
client 37 sending UPDATE pgbench_branches SET bbalance = bbalance + 4605 WHERE bid = 7;
client 37 receiving
client 37 receiving
client 37 sending INSERT INTO pgbench_history (tid, bid, aid, delta, mtime) VALUES (20, 7, 560242
, 4605, CURRENT_TIMESTAMP);
client 37 receiving
client 37 receiving
client 37 sending END;
client 37 receiving
client 37 receiving
transaction type: <builtin: TPC-B (sort of)>
scaling factor: 10
query mode: simple
number of clients: 150
number of threads: 2
number of transactions per client: 100
number of transactions actually processed: 7500/15000
latency average = 3138.897 ms
tps = 47.787488 (including connections establishing)
tps = 47.853332 (excluding connections establishing)
vagrant@devopsubuntu1804: $
```

Figure 10.5. Errors of a higher load than the PostgreSQL max_connections allowance

We can check the preceding figure and see that pgbench requires 150 users x 100 transactions = 15,000 transactions, but this load test is only able to complete 7,500 transactions because the maximum connections with PostgreSQL is only 87. Hence, we conclude that our RDS performance is not good for 150 users.

Because 150 users is a small number, we will now move on to the next section to resolve the preceding error so that our RDS can improve the performance. In addition to that error, pgbench statements always ask us to enter the RDS password, which we can utilize to resolve errors in the next section.

pgbench 3 – Creating and testing a connection pool

At the moment, Amazon Web Services is doing an experiment with a database connection pool service called RDS Proxy. In order to access the RDS Proxy service, please open the RDS dashboard (`https://console.aws.amazon.com/rds`) and select the **Proxies** tab on the left panel as shown in the following figure:

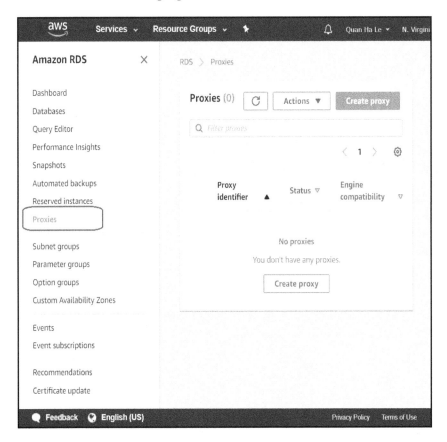

Figure 10.6. Experimental AWS RDS Proxy

Figure 10.7 shows how the RDS proxy works as a connection pool between applications and RDS databases:

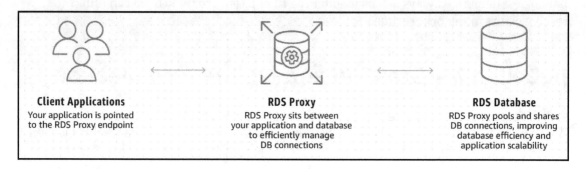

Figure 10.7. Principles of AWS RDS Proxy

However, currently, the RDS Proxy service is still in preview only. With this preview status, the RDS proxy service is able to create connection pools for Aurora MySQL RDS, MySQL RDS, and Aurora PostgreSQL RDS. This means that this AWS connection pool service is not available for our current PostgreSQL version 12.2.

Therefore, we are going to install a PostgreSQL connection pool on our own with either of the following:

- pgpool
- pgbouncer

I have already tried pgpool before, hence if the pgbench performance tool creates a higher number of user connections than the *max_connections = 87* allowance of our RDS, our test could show that pgpool uses the 2 parameters max_pool and num_init_children. The max_pool parameter configures how many connections pgpool can cache for each child process. For example, if num_init_children is configured to 100, and max_pool is configured to 3, then pgpool could potentially open 300 (=3*100) connections to the RDS.

However, pgpool also sets a restriction on the max_connections value of PostgreSQL RDS as you can see in the next formula:

```
max_pool*num_init_children <= (max_connections -
superuser_reserved_connections)
```

Our PostgreSQL RDS superuser_reserved_connections is 10.

Therefore, because of the preceding `pgpool` restriction, we cannot use `pgpool` to resolve our PostgreSQL connection error. Therefore, we will look towards `pgbouncer` and take the next steps:

1. We will install `pgbouncer` for the Jenkins server:

    ```
    vagrant@devopsubuntu1804:~$ sudo apt update
    vagrant@devopsubuntu1804:~$ echo "deb
    http://apt.postgresql.org/pub/repos/apt/ `lsb_release -cs`-pgdg
    main" | sudo tee  /etc/apt/sources.list.d/pgdg.list
    vagrant@devopsubuntu1804:~$ wget --quiet -O -
    https://www.postgresql.org/media/keys/ACCC4CF8.asc | sudo apt-key
    add -
    vagrant@devopsubuntu1804:~$ sudo apt update
    vagrant@devopsubuntu1804:~$ sudo apt install -y pgbouncer
    ```

The installation of `pgbouncer` is easy and smooth, as we saw previously:

Figure 10.8. Installation of pgbouncer

2. We will edit the configuration file for `pgbouncer` to connect to the RDS endpoint and the `atm` database. Only For demonstration purposes, we will use it *without ssl* and password authentication. Please do not use it in production environments:

```
vagrant@devopsubuntu1804:~$ sudo nano /etc/pgbouncer/pgbouncer.ini
------------------------------------------------------------------
--
[databases]
atm = host=atm.ck5074bwbilj.us-east-1.rds.amazonaws.com port=5432
dbname=atm

[pgbouncer]
listen_addr = *
listen_port = 6432
auth_type = trust
auth_file = /etc/pgbouncer/userlist.txt
pool_mode = transaction
max_client_conn = 200
max_db_connections = 70
server_reset_query =
------------------------------------------------------------------
--
```

Please use *Ctrl + O* to write the changes to the file:

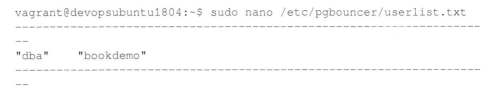

Figure 10.9. Configuration of pgbouncer

3. We will create the `pgbouncer` authorization file:

```
vagrant@devopsubuntu1804:~$ sudo nano /etc/pgbouncer/userlist.txt
-------------------------------------------------------------------
--
"dba"      "bookdemo"
-------------------------------------------------------------------
--
```

The user for `pgbouncer` to connect to the RDS is `dba` and the password is `bookdemo`.

4. At the end, we will restart the `pgbouncer` service:

```
vagrant@devopsubuntu1804:~$ sudo systemctl stop pgbouncer
vagrant@devopsubuntu1804:~$ sudo systemctl start pgbouncer
```

5. If all the settings are correct, we can call `psql` to connect from our Jenkins server to the RDS through port `6432` of pgbouncer:

```
vagrant@devopsubuntu1804:~$ sudo psql --dbname=atm --port=6432 --
username=dba
psql (12.2 (Ubuntu 12.2-2.pgdg18.04+1))
Type "help" for help.

atm=> SELECT * FROM public."ATM locations" LIMIT 10;
```

Hence, pgbouncer is working alright and it does a pooling task for the psql connection as follows:

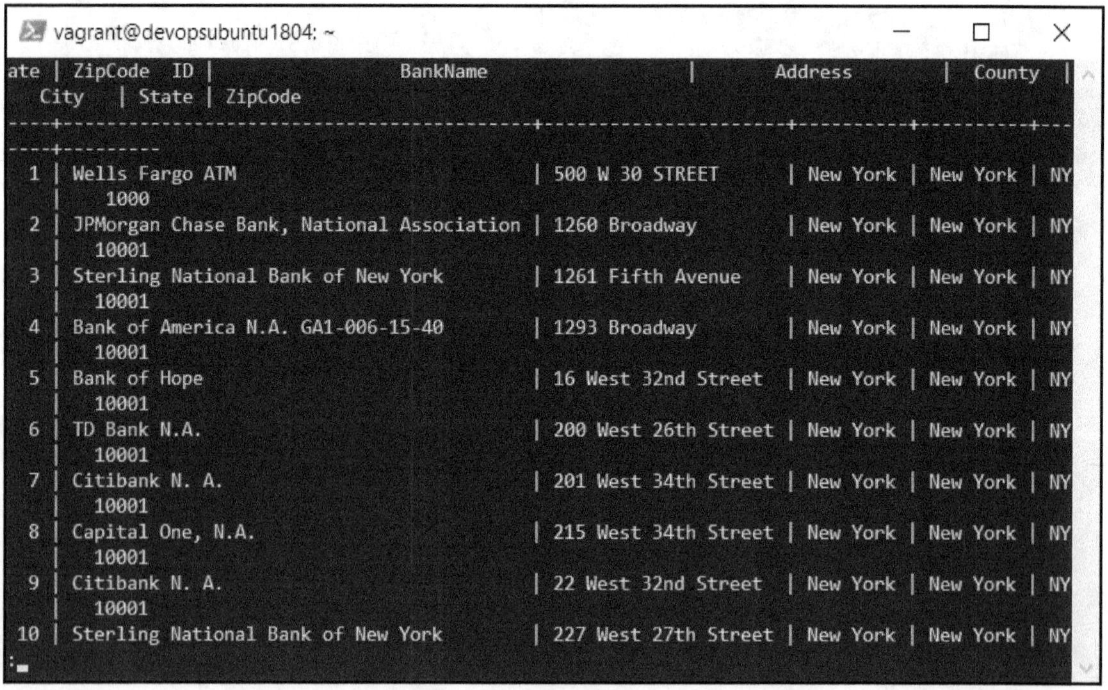

Figure 10.10. psql connection calls to pgbouncer port 6432

6. We'll now use pgbouncer port 6432 to create a pgbench load of 150 users more than the `max_connections` = 87 allowance of our RDS:

```
vagrant@devopsubuntu1804:~$ pgbench -h atm.ck5074bwbilj.us-
east-1.rds.amazonaws.com -p 5432 -U dba -c 150 -j 2 -t 100 -d atm

pghost: localhost pgport: 6432 nclients: 150 nxacts: 100 dbName:
atm
...
...
client 147 sending UPDATE pgbench_branches SET bbalance = bbalance
+ -4201 WHERE bid = 2;
client 147 receiving
client 148 receiving
client 110 receiving
client 110 sending INSERT INTO pgbench_history (tid, bid, aid,
delta, mtime) VALUES (29, 6, 647981, 3295, CURRENT_TIMESTAMP);
client 110 receiving
client 134 receiving
client 147 receiving
client 147 sending INSERT INTO pgbench_history (tid, bid, aid,
delta, mtime) VALUES (62, 2, 543082, -4201, CURRENT_TIMESTAMP);
client 147 receiving
client 110 receiving
client 110 sending END;
client 110 receiving
client 147 receiving
client 147 sending END;
client 147 receiving
client 110 receiving
client 147 receiving
transaction type: <builtin: TPC-B (sort of)>
scaling factor: 10
query mode: simple
number of clients: 150
number of threads: 2
number of transactions per client: 100
number of transactions actually processed: 15000/15000
latency average = 7540.436 ms
tps = 19.892748 (including connections establishing)
tps = 19.892775 (excluding connections establishing)
vagrant@devopsubuntu1804:~$
```

This time, pgbouncer makes a very good connection pool layer between pgbench and the RDS. We can see from the 150-user load that a total of 15,000 transactions are all completed with the throughput of roughly 20 completed transactions per second:

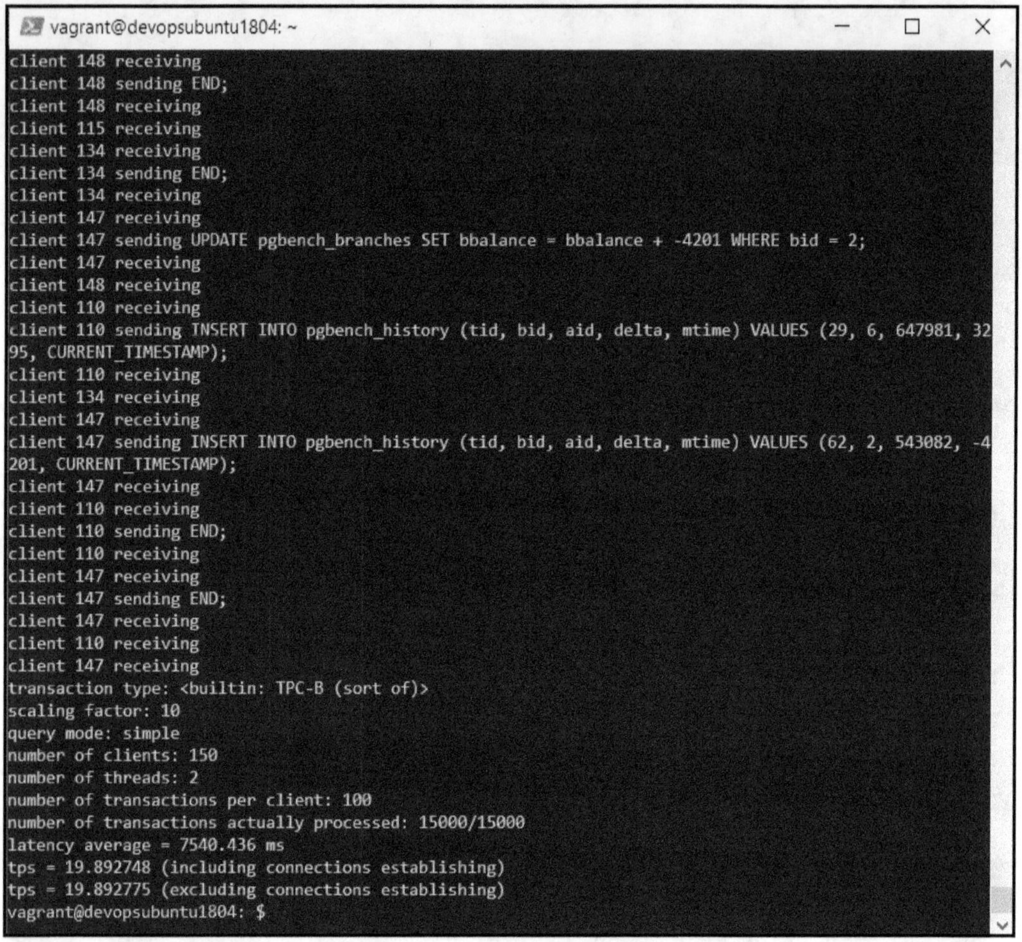

```
client 148 receiving
client 148 sending END;
client 148 receiving
client 115 receiving
client 134 receiving
client 134 sending END;
client 134 receiving
client 147 receiving
client 147 sending UPDATE pgbench_branches SET bbalance = bbalance + -4201 WHERE bid = 2;
client 147 receiving
client 148 receiving
client 110 receiving
client 110 sending INSERT INTO pgbench_history (tid, bid, aid, delta, mtime) VALUES (29, 6, 647981, 32
95, CURRENT_TIMESTAMP);
client 110 receiving
client 134 receiving
client 147 receiving
client 147 sending INSERT INTO pgbench_history (tid, bid, aid, delta, mtime) VALUES (62, 2, 543082, -4
201, CURRENT_TIMESTAMP);
client 147 receiving
client 110 receiving
client 110 sending END;
client 110 receiving
client 147 receiving
client 147 sending END;
client 147 receiving
client 110 receiving
client 147 receiving
transaction type: <builtin: TPC-B (sort of)>
scaling factor: 10
query mode: simple
number of clients: 150
number of threads: 2
number of transactions per client: 100
number of transactions actually processed: 15000/15000
latency average = 7540.436 ms
tps = 19.892748 (including connections establishing)
tps = 19.892775 (excluding connections establishing)
vagrant@devopsubuntu1804: $
```

Figure 10.11. pgbouncer makes a good connection pool for pgbench

Also, pgbouncer can connect automatically into the RDS without it asking for a password.

These are our settings for pgbouncer:

```
max_client_conn = 200
max_db_connections = 70
```

We can explain this result as follows:

- At first, pgbouncer accepts when pgbench launches 150 users because pgbouncer is set up with `max_client_conn` = 200, which is more than 150.
- Then pgbouncer pools 70 user connections (`max_db_connections` = 70) for pgbench to our RDS. This does not cause any errors because the allowance of PostgreSQL RDS is equal to
  `(max_connections - superuser_reserved_connections) = (87 - 10)`
  `= 77` connections larger than 70.
- Whenever each of the first 70 connections inside the pgbouncer pool has done its transactions, then pgbouncer can reuse the same idle connection for one of the remaining (150 - 70) users = 80 users.
- Therefore, pgbouncer creates a good connection pool to allow clients to execute a larger number of users than the real RDS connection allowance.

pgbouncer has done provided a good PostgreSQL benchmark but it generates its own data to send its own transactions over its pgbench tables only, therefore we decide to move on by using the JMeter tool so that we can use our atm data for load tests.

JMeter setup

Because JMeter will be easier to understand with the GUI, we are going to reuse the "QGIS Windows" EC2 instance from `Chapter 5`, *Creating a Geospatial Database Using PostGIS and PostgreSQL*, in the QGIS section.

We repeat the steps from Chapter 5 as follows:

1. Choose the **Amazon Machine Image (AMI)** from Chapter 5:

   Microsoft Windows Server 2019 Base – `ami-04ca2d0801450d495`
   This AMI can be seen at the top of *Figure 5.10* in Chapter 5.

2. The EC2 instance was shown in *Figure 5.11* in Chapter 5.
3. Please open the "QGIS Windows" EC2 instance with Remote Desktop Connection as shown in *Figure 5.12* in Chapter 5, Creating a Geospatial Database Using PostGIS and PostgreSQL.

3. Download and install the newest JMeter version here:
   `https://jmeter.apache.org/download_jmeter.cgi`

Apache JMeter 5.3 (Requires Java 8+)

Binaries

apache-jmeter-5.3.tgz sha512 pgp
apache-jmeter-5.3.zip sha512 pgp

Source

apache-jmeter-5.3_src.tgz sha512 pgp
apache-jmeter-5.3_src.zip sha512 pgp

Go to top

Figure 10.12. The newest JMeter versions

5. Let's save the JMeter ZIP file into a folder of EC2 such as `C:\JMeter\`. We then can extract the ZIP file into the folder `C:\JMeter\apache-jmeter-5.3\`.

6. We will then download the JDBC driver for PostgreSQL from the website `https://jdbc.postgresql.org/download.html`:

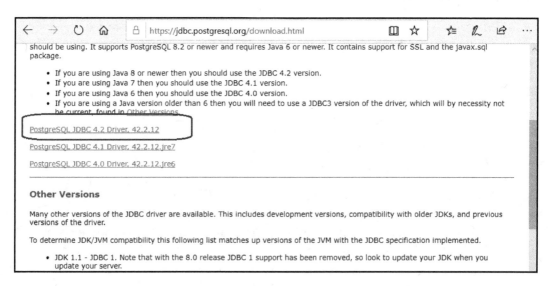

Figure 10.13. The PostgreSQL JDBC driver for JMeter

7. We have to save the JDBC driver file `postgresql-42.2.12.jar` to the JMeter lib folder, that is, `C:\JMeter\apache-jmeter-5.3\lib\`.

8. Now our JMeter is ready for PostgreSQL, we can execute the file `C:\JMeter\apache-jmeter-5.3\bin\jmeter.bat` to launch JMeter:

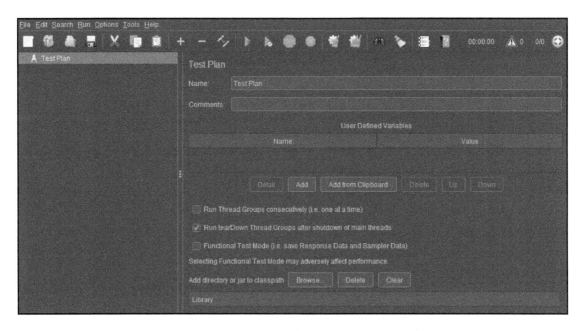

Figure 10.14. The first launch of Apache JMeter on Windows

In order to run a load test from JMeter, we will create a JMeter test plan in the next section.

JMeter for AWS PostgreSQL RDS

A JMeter test plan for AWS PostgreSQL RDS includes a user thread group; its thread group will include elements of variables, JDBC connections, JDBS requests, and performance reports:

1. Please create a thread group by using the menu **Edit** > **Add** > **Threads (Users)** > **Thread Group**. Or, we can simply right-click on **Test Plan** to pop up the **Add** > **Threads (Users)** > **Thread Group menu** options:

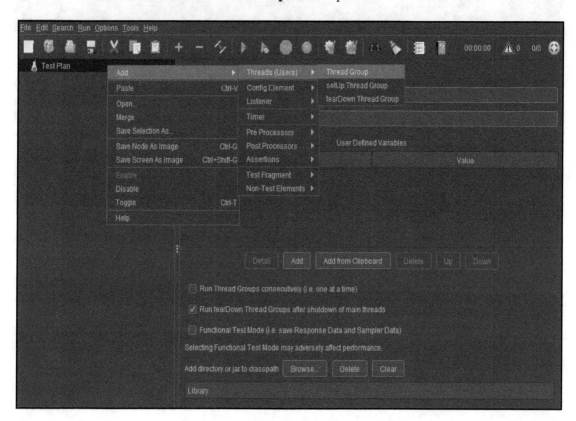

Figure 10.15. Creation of a new thread group by JMeter

We will see the default values of a thread group are as follows:

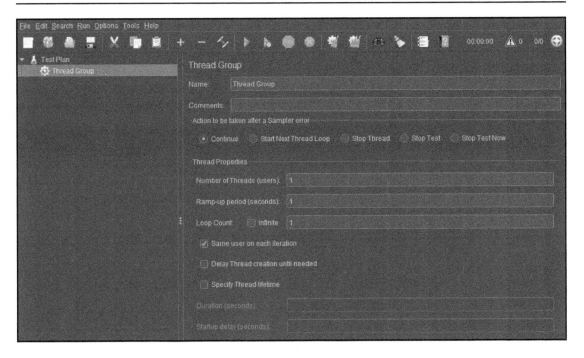

Figure 10.16. Thread Group default values

3. We will set up each value in **Thread Group** as follows:

- **Name:** PostgreSQL Users
- **Number of Threads (users):** 50
- **Ramp-up period (seconds):** 10
- **Loop Count:** 100
- To have JMeter repeatedly run your test plan, select the **Infinite** checkbox.

The **Ramp-up period** value tells JMeter how long to delay between starting each user. Now our ramp-up period is 10 seconds, JMeter has to start all of your users in 10 seconds. In this case, the number of users is 50, so the delay between starting users is equal to 10 seconds / 50 users = 0.2 seconds per user:

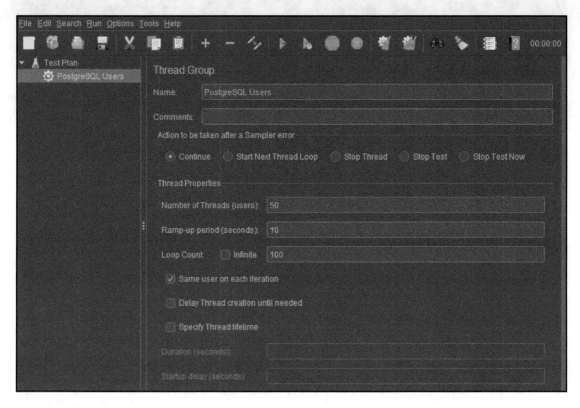

Figure 10.17. Number of users and Ramp-up period

4. We will add a random variable to generate a random value following the next steps:

First, right-click to get the contextual **Add** menu, and then select **Add** > **Config Element** > **Random Variable**. Then, select this new element to view its Control Panel:

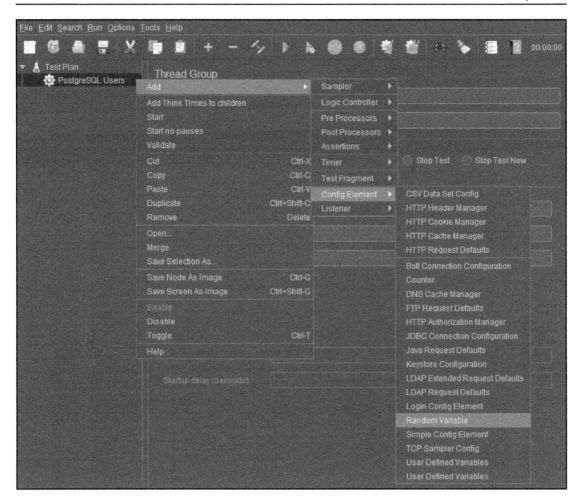

Figure 10.18. Adding a JMeter random variable

5. Once the variable is created, we will set up values for our random variable:

- **Name:** `Random ZipCode`
- **Variable Name:** `zip`
- **Minimum Value:** `10000`
- **Maximum Value:** `11011`

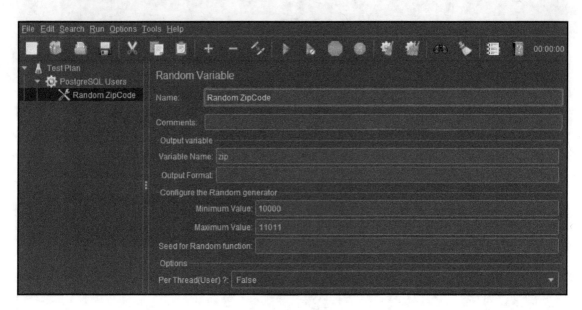

Figure 10.19. Random variable setup

6. We will set up a connection to the RDS on AWS.

Right-click to get the **Add** menu, and then select **Add** > **Config Element** > **JDBC Connection Configuration.** Then, select this new element to view its Control Panel:

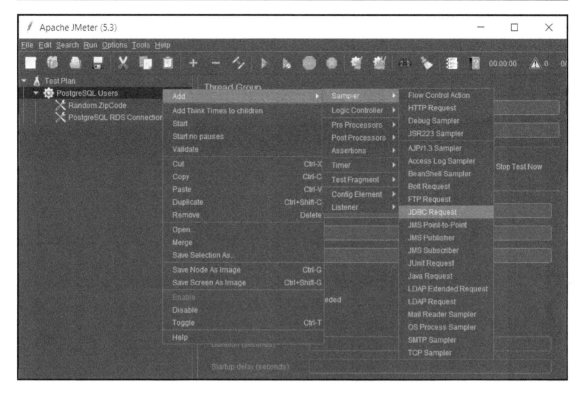

Figure 10.20. Adding a PostgreSQL connection

7. We will add values for the PostgreSQL connection:

- **Name**: `PostgreSQL RDS Connection`
- **Variable Name for created pool**: `ATM Database`
- **Database URL**: `jdbc:postgresql://atm.ck5074bwbilj.us-east-1.rds.amazonaws.com:5432/atm`

- **JDBC Driver class**: `org.postgresql.Driver`
- **Username**: `dba`
- **Password**: `bookdemo`

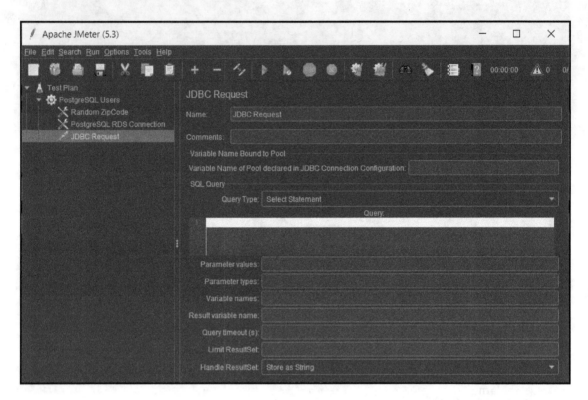

Figure 10.21. Connection details

8. We will add the first JDBC request.

   Click your right mouse button to get the **Add** menu, and then select **Add** > **Sampler** > **JDBC Request**. Then, select this new element to view its Control Panel:

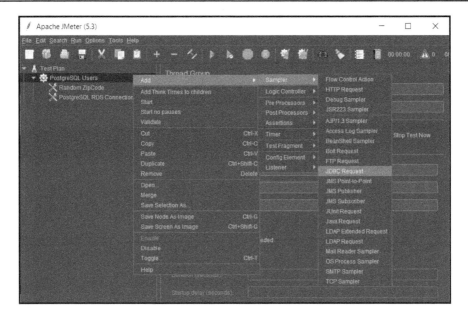

Figure 10.22. Adding a JDBC request

9. The **JDBC Request** form looks like this:

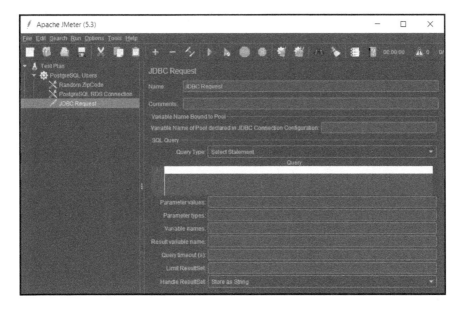

Figure 10.23. JDBC Request form

10. We will add our first `SELECT` query for the "ATM locations" table inside the RDS:

- Change **Name** to `ATM locations`.
- Enter the pool name: `ATM Database`.
- Enter the **SQL Query** string field: `SELECT "BankName", "Address", "County", "City", "State" FROM "ATM locations" WHERE "ZipCode" = ${zip};`

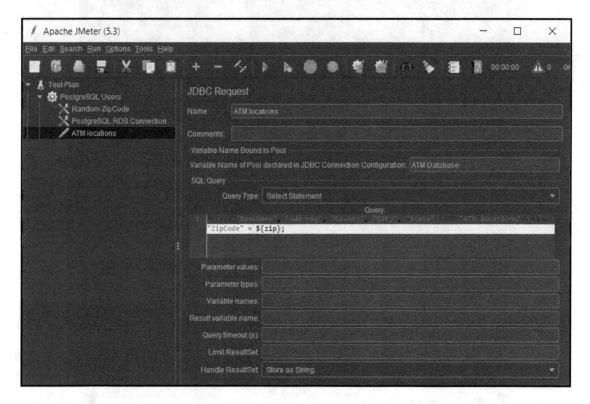

Figure 10.24. SELECT request

Our SELECT statement applies our random variable named **zip** that we set up in step 5.

11. Next, add the second JDBC request and edit the following properties:
    1. Change **Name** to `Zip coordinates`.
    2. Enter the pool name: `ATM Database`.
    3. Enter the **SQL Query** string field: `SELECT "city", "state", "geog" FROM "Zip coordinates" WHERE "zip" = ${zip};`

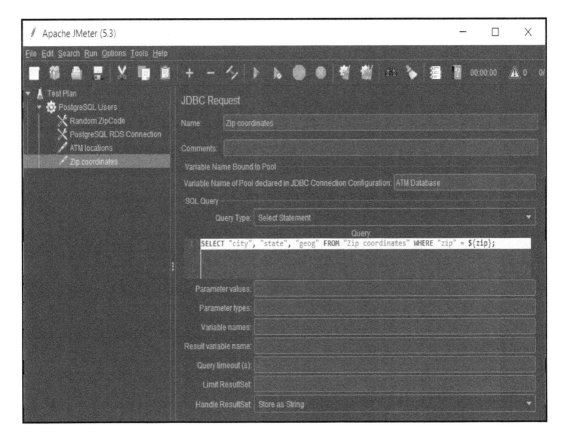

Figure 10.25. Second JDBC request

12. We will add two listeners to view/store the load test results:

First, we will add a view results tree (by using the menu **Add** > **Listener** > **View Results Tree**):

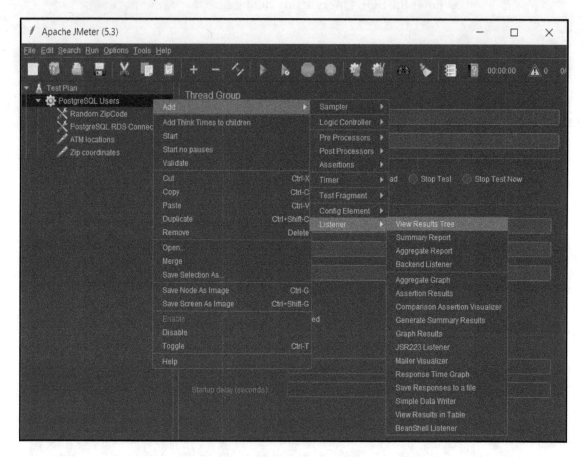

Figure 10.26. The View Results Tree listener

13. Then, add a summary report listener via the **Add** > **Listener** > **Summary Report** menu options:

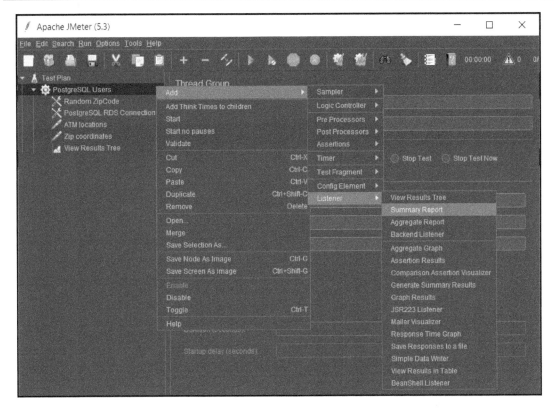

Figure 10.27. The Summary Report listener

14. Save the test plan by going to File -> Save or by clicking on the 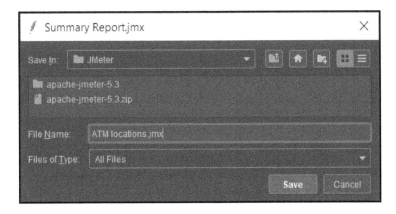 icon:

Figure 10.28. Saving the test plan file

15. Run the test with the menu **Run** > **Start** or *Ctrl + R*.

1. Select the **View Results Tree** tab.
2. Select the **Response data** tab.
3. Select **Response Body**.
4. Navigate down through each JDBC request.

You will be able to reach the retrieved data from the RDS:

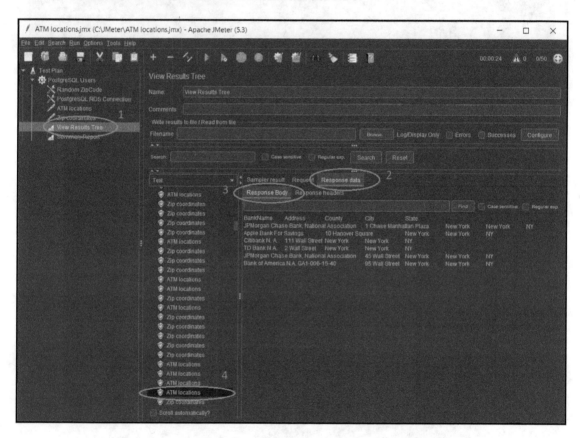

Figure 10.29. Checking each request execution with the View Results Tree listener

16. Select the **Summary Report** tab. You will be able to view the throughput. This time, we reached 404 JDBC requests per second:

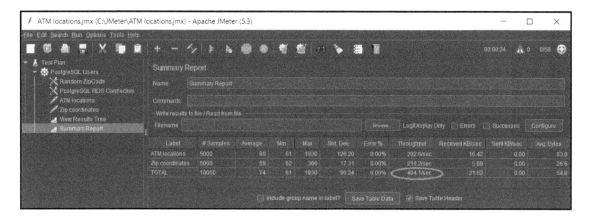

Figure 10.29. Final throughput in the Summary Report listener

JMeter is a good and stable tool. It is easy to learn and to use. Knowing JMeter gives system engineers the ability to complete load test requests for companies all over the world.

Summary

In this chapter, we have learned all about pgbench and how to configure its tests with pgbench parameters. We also learned about Pgbouncer and Pgpool, and then moved on to JMeter and JDBC configuration, threads, and listeners and also how to create JDBC requests, and how to use listener load reports.

In the next chapter, we will learn how to perform unit tests for PostgreSQL databases by using four test frameworks.

11
Test Frameworks for PostgreSQL

In this chapter, you will learn how to write automated tests for existing stored procedures or develop procedures using the concepts of unit tests and **Test-Driven Development** (**TDD**). All test cases are stored in the database, so we do not need any external resources (such as files, version control, and command-line utilities) to save tests. Database tests may be grouped into test cases; each test case may have its own environment initialization code. All tests are still executed independently because their effects are automatically rolled back after the execution.

With the help of the project in this chapter, we will set up and run automated unit tests to test whether possible mistakes exist in our PostgreSQL 12 RDS from **Amazon Web Services** (**AWS**) of Chapter 2, *Setting Up a PostgreSQL RDS for ATM Machines*, of our banking ATM machine locations within a typical city.

The following topics will be covered in the chapter:

- Making unit tests with pgTAP
- Making unit tests in a simpler way with PGUnit
- PGUnit – same name but a different approach
- Testing with Python – Tesgres

Technical requirements

This chapter will take developers around 16-20 hours of working to develop four PostgreSQL test frameworks.

Making unit tests with pgTAP

Through the different topics that we will develop here, we will be applying different techniques to carry out tests. We are going to start with one of the most used techniques in production environments: pgTAP, which has several functions for use case tests and also provides the output in TAP. The **Test Anything Protocol** (**TAP**) is well known for being suitable for harvesting, analysis, and reporting with a TAP harness.

The official website of pgTAP is `https://pgtap.org/`.

Setting up pgTAP for PostgreSQL RDS

We'll set up pgTAP using the following RDS:

1. Use pgAdmin to connect to our ATM RDS on AWS and select the ATM database to set up pgTAP.
2. Then, navigate to the top menu bar, go to **Tools** | **Query Tool**, and then execute the following SQL statements by pressing the ⚡ icon on the toolbar, as in the following figure, *Figure 11.1*:

```
CREATE SCHEMA IF NOT EXISTS pgTAP;
CREATE EXTENSION IF NOT EXISTS pgTAP WITH SCHEMA pgTAP;
```

This is better illustrated here:

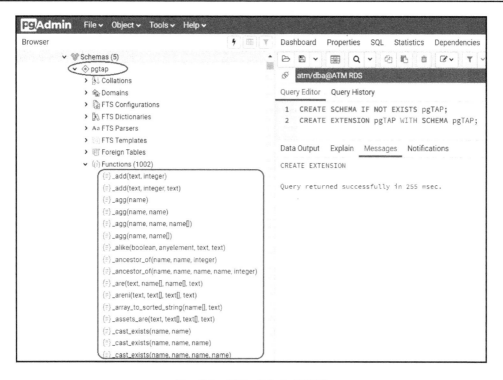

Figure 11.1 – pgTAP installation for AWS RDS

3. After that, you should be able to see the new pgTAP schema and the pgTAP stored procedures, as you can see in the preceding figure.

pgTAP test examples

Before anything else, you need a testing plan. This basically declares how many tests your script is going to run to protect against premature failure:

1. The preferred way to do this is to declare a plan by calling the plan () function for tests; if you only have one test, then you will call SELECT plan(1), but if you intend to implement *n* tests, you will declare SELECT plan(n):

```
SET search_path TO pgTAP;

SELECT plan(1);

SELECT is_empty('select "ID" from public."ATM locations" where
"ZipCode" NOT IN (select "zip" from public."Zip coordinates") LIMIT
```

```
1;', 'Check if there are any Bank ZipCodes not included in Zip
coordinates table');

SELECT * FROM finish();
```

2. The preceding SQL code is a test to see whether there are any bank ZIP codes inside the ATM locations table that are not found in the ZIP coordinates table, hence we execute that code from pgAdmin:

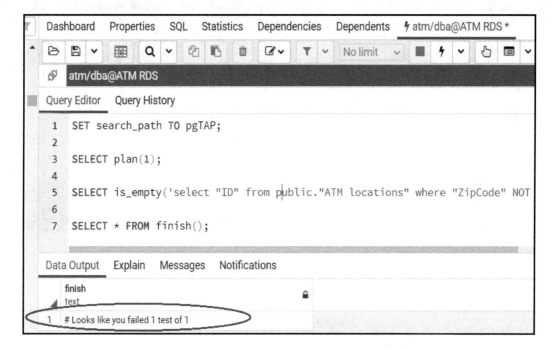

Figure 11.2 – pgTAP test 1

3. The output shows that the test fails because the first ATM has `ZipCode = 1000` but this value cannot be found in the ZIP coordinates table; the failure is `#` `Looks like you failed 1 test of 1`.

4. Now we test the second error, because the ATM database is used to store ATM locations of New York City, hence the ATM locations table must not be empty, so we have the following code for the next test:

```
SET search_path TO pgTAP;

SELECT plan(1);
```

```
SELECT isnt_empty('select distinct "ID" FROM public."ATM
locations";', 'Check if the ATM locations inside the ATM database
are not empty');

SELECT * FROM finish();
```

5. We now execute the second test with pgAdmin:

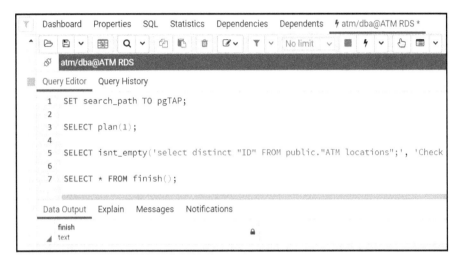

Figure 11.3 – pgTAP test 2

6. Our RDS passes the second test OK, which means our ATM locations table has 654 records and it is not empty.

7. Now, we are going to combine the two preceding tests together into one plan. So, our code will be designed for two tests with SELECT plan(2), as follows:

```
SET search_path TO pgTAP;

SELECT plan(2);

SELECT is_empty('select "ID" from public."ATM locations" where
"ZipCode" NOT IN (select "zip" from public."Zip coordinates") LIMIT
1;', 'Check if there are any Bank ZipCodes not included in Zip
coordinates table');

SELECT isnt_empty('select distinct "ID" FROM public."ATM
locations";', 'Check if the ATM locations inside the ATM database
are not empty');

SELECT * FROM finish();
```

8. Hence, when we execute the two-test plan with pgAdmin, the result will show #
   `Looks like you failed 1 test of 2`:

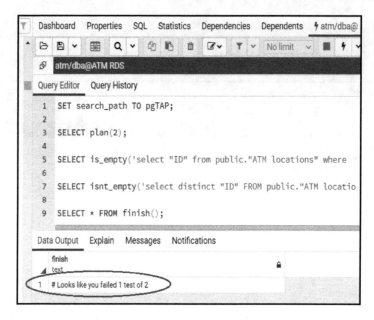

Figure 11.4 – The final pgTAP test plan

Uninstalling pgTAP for PostgreSQL RDS

In order to uninstall pgTAP, please use pgAdmin to execute the following two SQL statements:

```
DROP EXTENSION IF EXISTS pgTAP;
DROP SCHEMA IF EXISTS pgTAP;
```

Up to this point, we have seen one of the most important and updated test tools for PostgreSQL. It is easy to install and is available on different platforms such as Linux or Windows.

One of its greatest strengths lies in that TAP output is easily customizable to use in tools such as Jenkins for continuous integration.

Making unit tests in a simple way with PG_Unit

In the previous section, we have seen a tool that, thanks to its regular updates, has many functions to apply in test cases. Now we will continue with a tool that is unfortunately not so often updated but that, thanks to standardization and its simplicity, does not lose its validity: PG_Unit. We will see that when applying it in pgAdmin, we must make some fixes to the installation script but apart from that, we will be able to demonstrate a test case and thus have another option when carrying out test cases.

The website for PGUnit is here: `https://github.com/danblack/pgunit-sql/`.

Setting up PGUnit for PostgreSQL RDS

We'll start setting up pgunit for PostgreSQL using the following steps:

1. The SQL script to set up PGUnit is found here: `https://github.com/danblack/pgunit-sql/blob/master/sql/reinstall.sql`. Please open the preceding link in your browser. You will see the script shown in *Figure 11.5*:

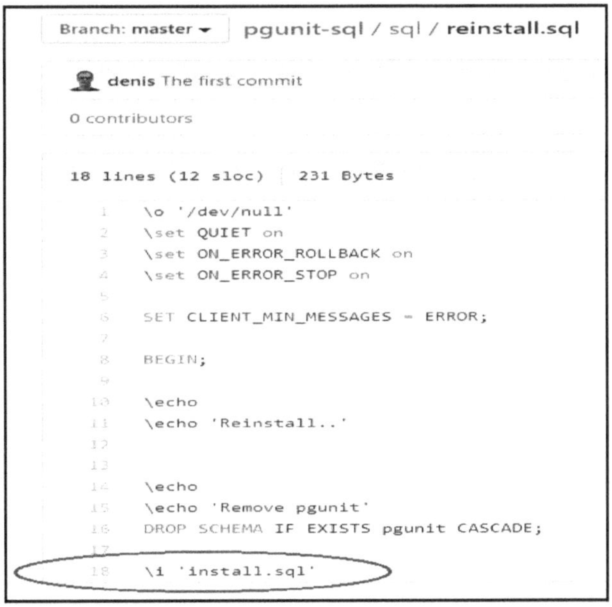

Figure 11.5 – PGUnit installation script

2. The `reinstall.sql` installation script calls to execute another `install.sql` file and in turn, the `install.sql` script calls to many other scrips, such as `assert_array_equals.sp.sql`, `assert_equals.sp.sql`, and `assert_false.sp.sql`, but we are not using `psql` as these scripts are intending:

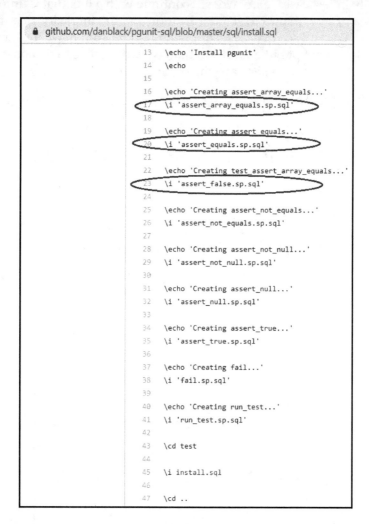

```
🔒 github.com/danblack/pgunit-sql/blob/master/sql/install.sql

13    \echo 'Install pgunit'
14    \echo
15
16    \echo 'Creating assert_array_equals...'
17    \i 'assert_array_equals.sp.sql'
18
19    \echo 'Creating assert equals...'
20    \i 'assert_equals.sp.sql'
21
22    \echo 'Creating test_assert_array_equals...'
23    \i 'assert_false.sp.sql'
24
25    \echo 'Creating assert_not_equals...'
26    \i 'assert_not_equals.sp.sql'
27
28    \echo 'Creating assert_not_null...'
29    \i 'assert_not_null.sp.sql'
30
31    \echo 'Creating assert_null...'
32    \i 'assert_null.sp.sql'
33
34    \echo 'Creating assert_true...'
35    \i 'assert_true.sp.sql'
36
37    \echo 'Creating fail...'
38    \i 'fail.sp.sql'
39
40    \echo 'Creating run_test...'
41    \i 'run_test.sp.sql'
42
43    \cd test
44
45    \i install.sql
46
47    \cd ..
```

Figure 11.6 – The inner install.sql script

3. Because we are using pgAdmin for the AWS PostgreSQL RDS and we are not using `psql` command-line scripts, we have to comment out `\o` and `\echo` for command-line `psql` command usages and then we have to copy thee sub-files' scripts to replace in the `reinstall.sql` script. Also, the following three statements need commenting out:

- `\set QUIET on`
- `\set ON_ERROR_ROLLBACK on`
- `\set ON_ERROR_STOP on`

4. We copy and execute the script in pgAdmin as in the following screenshot:

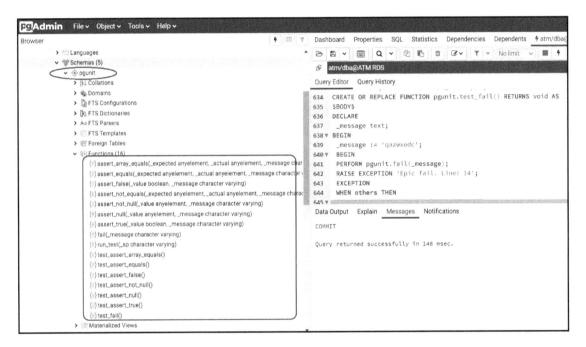

Figure 11.7 – Completion of PGUnit installation

5. The library is based on stored procedures that compare variables between themselves or compare variables with various constants:

- `pgunit. assert_equals (_expected anyelement, actual anyelement, custom_message varchar)`: Compares two elements. If they are not equal, then the `#assert_equalsn {custom_message}` exception is thrown.

- pgunit. assert_not_equals (_expected anyelement, actual anyelement, custom_message varchar): Compares two elements. If they are equal, an exception is thrown: #assert_not_equalsn {custom_message}.

- pgunit. assert_array_equals (_expected anyelement [ ], actual [] anyelement, custom_message varchar): Compares two arrays. Arrays are considered equal if these arrays have the same elements and the sizes of the arrays are equal. If the arrays are not equal, an exception is thrown with the text #assert_array_equalsn {custom_message}.

- pgunit. assert_true (_value boolean, _custom_message varchar): Compares _value to True. If they are not equal, a #assert_truen {custom_message} exception is thrown.

- pgunit. assert_false (_value boolean, _custom_message varchar): Compares _value to False. If they are not equal, an exception is thrown: #assert_falsen {custom_message}.

- pgunit. assert_null (_value boolean, _custom_message varchar): Compares _value to NULL. If they are not equal, a #assert_nulln {custom_message} exception is thrown.

- pgunit. assert_not_null (_value boolean, _custom_message varchar):
  Compares _value to NULL. If they are equal, an exception is thrown: #assert_not_nulln {custom_message}.

- pgunit. fail (_custom_message varchar): Throws an exception with the text #assert_failn {custom_message}.

- pgunit. run_test (_sp varchar): Runs the specified stored procedure inside the test infrastructure. After starting the test procedure, data is rolled back.

PGUnit test examples

We'll check out some pgunit test examples as follows:

1. This is the PGUnit script for the same first test of checking whether there are any bank ATM ZIP codes that are not included in the ZIP coordinate table:

```
create or replace function pgunit.__test_bank_zipcode() returns
void as $$
declare
    id INT;
begin
    select "ATM locations"."ID" from "ATM locations" where "ATM
```

```
locations"."ZipCode" NOT IN (select "Zip coordinates".zip from "Zip
coordinates") LIMIT 1 INTO id;
    perform pgunit.assert_null(id, 'Bank ZipCode is not included in
Zip coordinates table');
end;
$$ language plpgsql;
```

2. After creating the first test in pgAdmin, we can execute the following query to run the first test:

```
select * from pgunit.__test_bank_zipcode();
```

3. Because the first ATM location has `zipcode = 1000`, which does not exist in the ZIP coordinates table, an exception is raised with our defined message: `Bank ZipCode is not included in Zip coordinates table`:

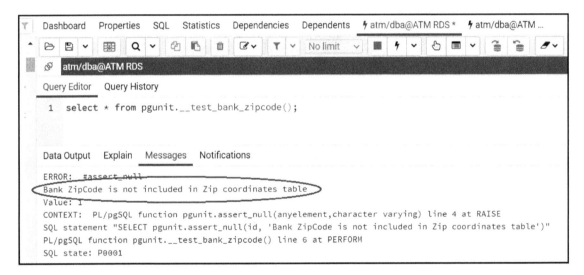

Figure 11.8 – PGUnit test 1

4. This is the PGUnit script for the same second test of checking that the ATM locations should not be empty:

```
create or replace function pgunit.__test_atm_locations() returns
void as $$
declare
    atmtotal INT;
begin
    SELECT count(distinct "ID") INTO atmtotal FROM public."ATM
locations";
```

```
        PERFORM pgunit.assert_true ((atmtotal IS NOT NULL) AND
    (atmtotal > 0), 'There are no ATM locations inside the ATM
    database');
    end;
    $$ language plpgsql;
```

5. After creating the second test in pgAdmin, we can execute the following query to run the second test:

```
select * from pgunit.__test_atm_locations();
```

6. Because we already stored 654 ATM locations, the second test should pass alright with no exceptions:

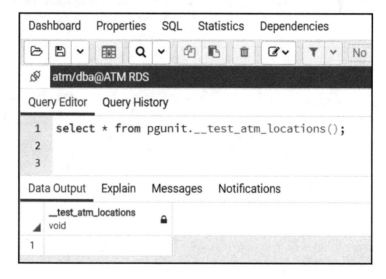

Figure 11.9 – PGUnit test 2

Uninstalling PGUnit for PostgreSQL RDS

In order to uninstall PGUnit, please use pgAdmin to execute the following SQL statement:

```
DROP SCHEMA IF EXISTS pgunit CASCADE;
//
```

As we have seen, PG_Unit is a simple tool for PostgreSQL tests; while it's not full of features for testing, it works in a modest way and we have verified it in the preceding sections.

PGUnit – same name but a different approach

When we decided to apply unit tests in PostgreSQL, we noticed there were two tools with the same name: PGUnit, but when we tried them, we realized that they share the fact of simplifying the tests but, unlike the previous one, the pgunit we are using has more recent updates. In the following steps, we will take care of applying it to our project and evaluating its behavior.

The website for simple pgunit is found here: `https://github.com/adrianandrei-ca/pgunit`.

Setting up simple pgunit for PostgreSQL RDS

1. The `plpgsql` code depends on the `dblink` extension being present in the database you run the tests on. We set up a simple pgunit in a dedicated schema such as `pgunit` and run these two lines of SQL:

   ```
   CREATE SCHEMA pgunit;
   CREATE EXTENSION DBLINK SCHEMA pgunit;
   ```

2. You should run the `PGUnit.sql` code using pgAdmin: `https://github.com/adrianandrei-ca/pgunit/blob/master/PGUnit.sql`.

3. We will copy the code from GitHub:

```
339 lines (317 sloc)    11 KB
 1
 2    create type test_results as (
 3      test_name varchar,
 4      successful boolean,
 5      failed boolean,
 6      erroneous boolean,
 7      error_message varchar,
 8      duration interval);
 9
10    --
11    -- Use select * from test_run_all() to execute all test cases
12    --
13    create or replace function test_run_all() returns setof test_results as $$
14    begin
15      return query select * from test_run_suite(NULL);
16    end;
17    $$ language plpgsql set search_path from current;
18
19    --
20    -- Executes all test cases part of a suite and returns the test results.
21    --
22    -- Each test case will have a setup procedure run first, then a precondition,
23    -- then the test itself, followed by a postcondition and a tear down.
24    --
25    -- The test case stored procedure name has to match 'test_case_<p_suite>%' patern.
26    -- It is assumed the setup and precondition procedures are in the same schema as
27    -- the test stored procedure.
28    --
29    -- select * from test_run_suite('my_test'); will run all tests that will have
30    -- 'test_case_my_test' prefix.
31    create or replace function test_run_suite(p_suite TEXT) returns setof test_results as $$
32    declare
33      l_proc RECORD;
34      l_sid INTEGER;
```

Figure 11.10 – Simple pgunit installation script

4. A convenient way to install the simple pgunit suite in our dedicated pgunit schema is to temporarily change the search path like this:

```
SET search_path TO pgunit;
```

We then revert search_path back to the public schema after the GitHub script.

5. Thereafter, please execute the whole of the script together in pgAdmin.

6. The result of a simple pgunit should be as follows:

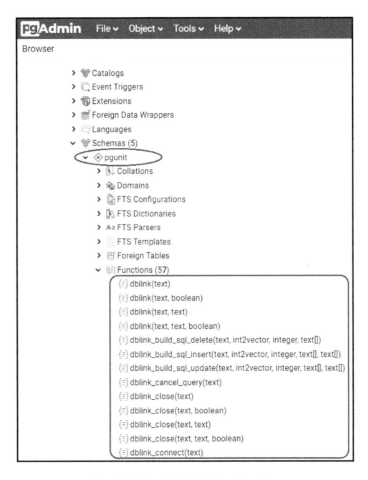

Figure 11.11 – Completion of the simple pgunit installation

Here is a list of prefixes for all tests:

- `"test_case_"`: It is a unit test procedure.
- `"test_precondition_"`: It is a test precondition function.
- `"test_postcondition_"`: It is a test postcondition function.
- `"test_setup_"`: It is a test setup procedure.
- `"test_teardown_"`: It is a test tear-down procedure.

Simple pgunit test examples

We'll now check out some pgunit test examples as follows:

1. This is the first test script to test whether there are any bank ATM ZIP codes not included in the ZIP coordinates table:

```
create or replace function pgunit.test_case_bank_zipcode() returns
void as $$
declare
    id INT;
begin
    select "ATM locations"."ID" from "ATM locations" where "ATM
locations"."ZipCode" NOT IN (select "Zip coordinates".zip from "Zip
coordinates") LIMIT 1 INTO id;
    perform pgunit.test_assertNull('Bank ZipCode is not included in
Zip coordinates table', id); end;
$$ language plpgsql;
```

2. After creating the first test by pgAdmin, please execute this query to proceed with the first test:

```
select * from pgunit.test_case_bank_zipcode();
```

3. The result is as follows:

Figure 11.12 – Simple pgunit test 1

The first bank ATM location has `zipcode = 1000` but as this value, `1000`, does not exist in the ZIP coordinates table, the `Bank ZipCode is not included in Zip coordinates table` exception is raised.

4. The second test to make sure that the ATM locations table is not empty is a precondition function:

```
create or replace function pgunit.test_precondition_atm_locations()
returns void as $$
declare
    atmtotal INT;
begin
    SELECT count(distinct "ID") INTO atmtotal FROM public."ATM
locations";
    PERFORM pgunit.test_assertTrue('There are no ATM locations
inside the ATM database', (atmtotal IS NOT NULL) AND (atmtotal > 0)
);
end;
$$ language plpgsql;
```

5. After creating the first test by pgAdmin, please execute this query to proceed with the second test:

```
select * from pgunit.test_precondition_atm_locations();
```

6. The result of the second test is shown:

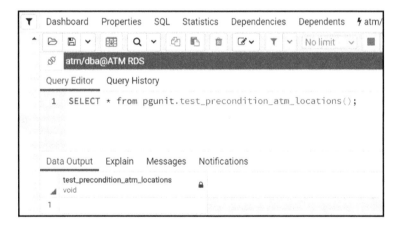

Figure 11.13 – Simple pgunit test 2

There are no exceptions raised because our ATM locations table has 654 records and it is not empty.

For general tests of simple pgunit, such as `select * from pgunit.test_run_all()`, because AWS does not support the `dblink_connect_u` function on RDS for PostgreSQL, these general tests are not supported in AWS RDS.

Uninstalling simple pgunit for PostgreSQL RDS

We can uninstall simple pgunit using the following steps:

1. `PGUnitDrop.sql` has the code you can use to remove all PGUnit code from the database: `https://github.com/adrianandrei-ca/pgunit/blob/master/PGUnitDrop.sql`.

2. We will copy the content from the browser:

Figure 11.14 – Un-installation of the simple pgunit

3. Because we have installed all of the simple pgunit library inside our separate `pgunit` schema, we modify the preceding script to look like the following:

```
--
-- Clears the PG Unit functions
--
--
drop function pgunit.test_run_suite(TEXT);
drop function pgunit.test_run_all();
drop function pgunit.test_run_condition(proc_name text);
drop function pgunit.test_build_procname(parts text[], p_from
integer, p_to integer);
drop function pgunit.test_get_procname(test_case_name text,
expected_name_count integer, result_prefix text);
drop function pgunit.test_terminate(db VARCHAR);
drop function pgunit.test_autonomous(p_statement VARCHAR);
drop function pgunit.test_dblink_connect(text, text);
drop function pgunit.test_dblink_disconnect(text);
drop function pgunit.test_dblink_exec(text, text);
drop function pgunit.test_detect_dblink_schema();
drop function pgunit.test_assertTrue(message VARCHAR, condition
BOOLEAN);
drop function pgunit.test_assertTrue(condition BOOLEAN);
drop function pgunit.test_assertNotNull(VARCHAR, ANYELEMENT);
drop function pgunit.test_assertNull(VARCHAR, ANYELEMENT);
drop function pgunit.test_fail(VARCHAR);
drop type pgunit.test_results cascade;

DROP EXTENSION IF EXISTS DBLINK;
DROP SCHEMA IF EXISTS pgunit CASCADE;
```

We will execute the preceding script in pgAdmin to uninstall the simple pgunit.

As we have seen, using pgunit is pretty straightforward to install but at times uninstalling it is a bit more complicated. Despite this, we were able to run a case test successfully.

In the next section, we will work with another test approach based on Python.

Testing with Python – Testgres

Testgres was developed under the influence of the Postgres TAP test feature. As an extra feature, it can manage Postgres clusters: initialize, edit configuration files, start/stop the cluster, and execute queries. In the following steps, we will see how to install it, execute a case test, and properly uninstall it.

The website for Testgres is found here: `https://github.com/postgrespro/testgres`.

Setting up Testgres for PostgreSQL

Testgres is a Python test tool, hence we will set up Testgres on our Jenkins Ubuntu server `192.168.0.200` that we set up with Vagrant in Chapter 7, *PostgreSQL with DevOps for Continuous Delivery*, to connect to our RDS on AWS:

1. We open SSH into the Jenkins server. Please launch PowerShell as an administrator:

   ```
   PS C:\Windows\system32>
   PS C:\Windows\system32> cd C:\Projects\Vagrant\Jenkins
   PS C:\Projects\Vagrant\Jenkins> vagrant up --provider virtualbox
   PS C:\Projects\Vagrant\Jenkins> vagrant ssh

   vagrant@devopsubuntu1804:~$
   ```

2. Install `pip3`:

   ```
   vagrant@devopsubuntu1804:~$ sudo apt install -y python3-pip
   ```

3. Please answer `<Yes>` when you get to the **Package configuration** screen:

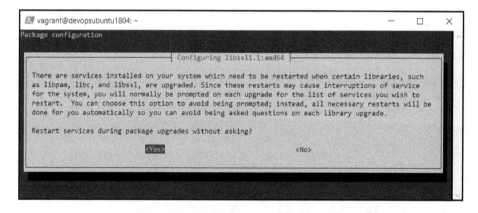

Figure 11.15 – Package configuration screen

4. You can check the version of `pip3`:

```
vagrant@devopsubuntu1804:~$ pip3 -V
pip 9.0.1 from /usr/lib/python3/dist-packages (python 3.6)
```

5. Install Testgres using `pip3`:

```
vagrant@devopsubuntu1804:~$ sudo pip3 install testgres
```

This is a screenshot of the Testgres installation:

Figure 11.16 – Testgres installation

7. We complete the Testgres installation by setting the `PG_BIN` and `PYTHON_VERSION` environment variables and then storing them:

```
vagrant@devopsubuntu1804:~$ sudo su
root@devopsubuntu1804:~# export PG_BIN=/usr/lib/postgresql/12/bin
root@devopsubuntu1804:~# export PYTHON_VERSION=3
root@devopsubuntu1804:~# source ~/.profile
```

Testgres test examples

We will check out some examples from Testgres:

1. We create the same two tests for our RDS, as follows:

```
root@devopsubuntu1804:~# mkdir /usr/local/src/testgres
root@devopsubuntu1804:~# nano /usr/local/src/testgres/atmrds.py
--------------------------------------------------------------
#!/usr/bin/env python
# coding: utf-8

import unittest
import logging
import testgres

logging.basicConfig(filename='/tmp/testgres.log')

class RdsTest(unittest.TestCase):
    def test_bank_zipcode(self):
        #create a node with random name, port, etc
        with testgres.get_new_node() as node:
            # RDS endpoint
            node.host = 'atm.ck5074bwbilj.us-
east-1.rds.amazonaws.com'
            node.port = 5432

            # run inidb
            node.init()

        with node.connect('atm', 'dba', 'bookdemo') as conn:
            conn.begin('serializable')

            # execute a query in the AWS RDS
            data = conn.execute('select "ATM locations"."ID"
from "ATM locations" where "ATM locations"."ZipCode" NOT IN (select
"Zip coordinates".zip from "Zip coordinates") LIMIT 1;')
            self.assertFalse(data, 'Bank ZipCode is not included
in Zip coordinates table')

    def test_atm_locations(self):
        #create a node with random name, port, etc
        with testgres.get_new_node() as node:
            # RDS endpoint
            node.host = 'atm.ck5074bwbilj.us-
east-1.rds.amazonaws.com'
            node.port = 5432
```

```
# run inidb
node.init()

with node.connect('atm', 'dba', 'bookdemo') as conn:
    conn.begin('serializable')

    # execute a query in the AWS RDS
    data = conn.execute('SELECT count(distinct "ID")
FROM public."ATM locations";')
        self.assertTrue(data, 'There are no ATM locations
inside the ATM database')
        self.assertGreater(data[0][0], 0, 'There are no ATM
locations inside the ATM database')

if __name__ == '__main__':
    unittest.main()
```

The test_bank_zipcode function is the first test, to check whether there are any bank ATM ZIP codes that not included in the ZIP coordinates table, and the second test_atm_locations function checks that the ATM locations table is not empty.

2. Please set the 755 permission for the test:

```
root@devopsubuntu1804:~# chmod 755
/usr/local/src/testgres/atmrds.py
```

3. With the correct permission, we can execute the test for our RDS as a Postgres user:

```
root@devopsubuntu1804:~# cd /usr/local/src/testgres/
root@devopsubuntu1804:/usr/local/src/testgres# su postgres -c
'python3 atmrds.py'
.F
=======================================================================
===
FAIL: test_bank_zipcode (__main__.RdsTest)
-----------------------------------------------------------------------
---
Traceback (most recent call last):
  File "atmrds.py", line 26, in test_bank_zipcode
    self.assertFalse(data, 'Bank ZipCode is not included in Zip
coordinates table')
AssertionError: [(1,)] is not false : Bank ZipCode is not included
in Zip coordinates table
```

```
------------------------------------------------------------------
---
Ran 2 tests in 1.941s

FAILED (failures=1)
```

We can see this is in the screenshot here:

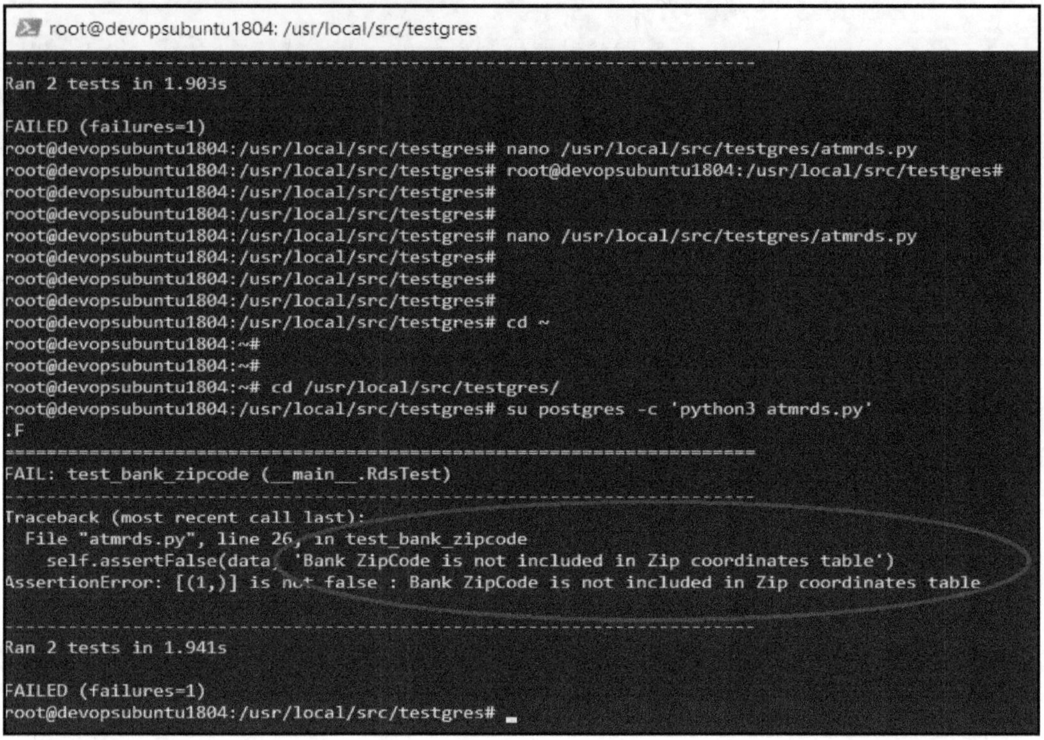

Figure 11.17 – Testgres tests

The two tests return one failure with the exception as `Bank ZipCode is not included in Zip coordinates table`.

5. *Steps 1-4* show how to run tests directly with `python3`. We can improve our testing by using `virtualenv`. Please install `virtualenv`:

```
root@devopsubuntu1804:/usr/local/src/testgres# cd /root
root@devopsubuntu1804:~# pip3 install virtualenv
```

6. A screenshot of the `virtualenv` installation is as follows:

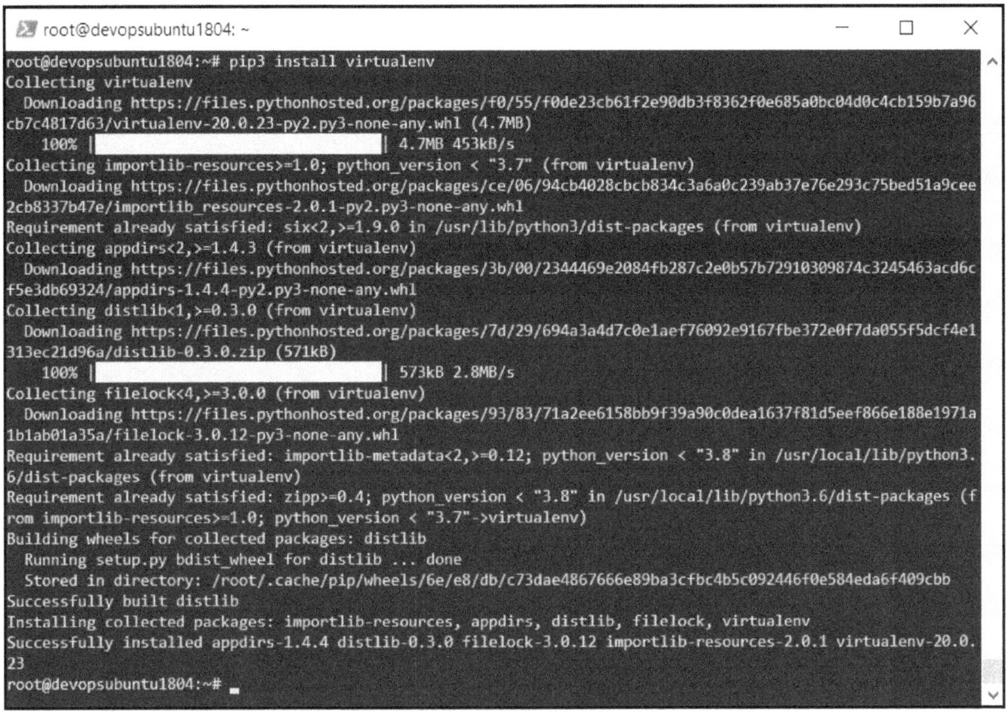

Figure 11.18 – virtualenv installation

7. Now we can create the Bash script for `virtualenv` as follows. This script calls to the `python3 '/usr/local/src/testgres/atmrds.py'` test file:

```
root@devopsubuntu1804:~# nano /usr/local/src/testgres/rds_tests.sh
-----------------------------------------------------------------
#!/usr/bin/env bash

VIRTUALENV="virtualenv --python=/usr/bin/python$PYTHON_VERSION"
PIP="pip$PYTHON_VERSION"

# prepare environment
VENV_PATH=/tmp/testgres_venv
rm -rf $VENV_PATH
$VIRTUALENV $VENV_PATH
export VIRTUAL_ENV_DISABLE_PROMPT=1
source $VENV_PATH/bin/activate

# Install local version of testgres
$PIP install testgres
```

```
# run tests (PG_BIN)
/usr/local/src/testgres/atmrds.py
------------------------------------------------------------------
```

8. Please grant execution permission for the script:

```
root@devopsubuntu1804:~# chmod 755
/usr/local/src/testgres/rds_tests.sh
```

9. We navigate away from the current /root folder and then execute the virtualenv Bash script as a Postgres user:

```
root@devopsubuntu1804:~# cd /usr/local/src/testgres/

root@devopsubuntu1804:/usr/local/src/testgres# su postgres -c
'./rds_tests.sh'
created virtual environment CPython3.6.9.final.0-64 in 184ms
  creator CPython3Posix(dest=/tmp/testgres_venv, clear=False,
global=False)
  seeder FromAppData(download=False, pip=latest, setuptools=latest,
wheel=latest, via=copy,
app_data_dir=/var/lib/postgresql/.local/share/virtualenv/seed-app-
data/v1.0.1)
  activators
BashActivator,CShellActivator,FishActivator,PowerShellActivator,Pyt
honActivator,XonshActivator
Processing
/var/lib/postgresql/.cache/pip/wheels/a7/2c/a6/37870923d4e356392e53
abab3a242cc67535075480e6c177b0/testgres-1.8.3-py3-none-any.whl
Collecting pg8000
  Using cached pg8000-1.15.3-py3-none-any.whl (24 kB)
Processing
/var/lib/postgresql/.cache/pip/wheels/a1/d9/f2/b5620c01e9b3e858c687
7b1045fda5b115cf7df6490f883382/psutil-5.7.0-cp36-cp36m-
linux_x86_64.whl
Collecting port-for>=0.4
  Using cached port_for-0.4-py2.py3-none-any.whl (21 kB)
Collecting six>=1.9.0
  Using cached six-1.15.0-py2.py3-none-any.whl (10 kB)
Collecting scramp==1.2.0
  Using cached scramp-1.2.0-py3-none-any.whl (6.3 kB)
Installing collected packages: scramp, pg8000, psutil, port-for,
six, testgres
Successfully installed pg8000-1.15.3 port-for-0.4 psutil-5.7.0
scramp-1.2.0 six-1.15.0 testgres-1.8.3
.F
======================================================================
===
```

```
FAIL: test_bank_zipcode (__main__.RdsTest)
----------------------------------------------------------------
---
Traceback (most recent call last):
  File "/usr/local/src/testgres/atmrds.py", line 26, in
test_bank_zipcode
    self.assertFalse(data, 'Bank ZipCode is not included in Zip
coordinates table')
AssertionError: [(1,)] is not false : Bank ZipCode is not included
in Zip coordinates table


----------------------------------------------------------------
---
Ran 2 tests in 2.018s

FAILED (failures=1)
root@devopsubuntu1804:/usr/local/src/testgres#
```

10. The result confirms that we execute two tests and the first test has failed with the Bank ZipCode is not included in Zip coordinates table exception message:

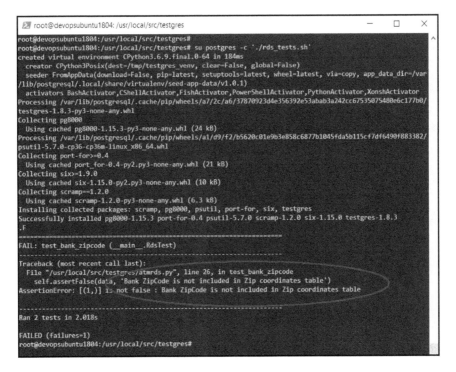

Figure 11.19 – Running Testgres tests with virtualenv

Uninstalling Testgres for PostgreSQL RDS

Because we have installed Testgres with the Vagrant user, now we can exit from the root user to perform Testgres uninstallation as follows:

```
root@devopsubuntu1804:/usr/local/src/testgres# exit
vagrant@devopsubuntu1804:~$ sudo -H pip3 uninstall -y testgres
Uninstalling testgres-1.8.3:
  Successfully uninstalled testgres-1.8.3
```

As we have seen, Testgres is a serious test framework. It should be taken into consideration when testing in a real production environment. A disadvantage could be that it is hard to install in a Windows environment.

Summary

In this chapter, with the help of a step-by-step project, we have learned PostgreSQL test skills such as schema validation and xUnit-style testing through pgTAP, automated tests for existing stored procedures or developing procedures with PGUnit and simple pgunit, and PostgreSQL node control with Testgres.

This book is now complete. We can proceed on to the book's *Appendix* to learn about PostgreSQL using other clouds that are different from AWS.

12

APPENDIX - PostgreSQL among the Other Current Clouds

In this chapter, we will learn about the **Database as a Service** (**DBaaS**) options for PostgreSQL with other cloud platforms, such as Google Cloud SQL for PostgreSQL, Heroku PostgreSQL, **EnterpriseDB** (**EDB**) cloud database, Microsoft Azure for PostgreSQL, and more.

The projects of this *Appendix* will set up PostgreSQL 12 to store ATM locations within a typical city for banking systems on Google Cloud SQL, Microsoft Azure, Heroku, and the EDB cloud.

We will cover the following topics in this chapter:

- Google Cloud SQL
- Microsoft Azure for PostgreSQL
- Heroku Postgre
- EnterpriseDB cloud database

Technical requirements

This *Appendix* will take developers around 20-22 hours to develop PostgreSQL databases on Google Cloud SQL, Microsoft Azure, Heroku, and the EDB cloud.

Google Cloud SQL

Google Cloud SQL is a high-performing managed database compared to PostgreSQL and others engines from Google.

As with most cloud services, it is completely managed through a simple interface, integrated with the other Google Cloud services, and perhaps most importantly, is reliable due to configuring replication and backups with ease to protect data.

Creating an instance

In the following steps, we will set up an instance through Cloud Console. Let's see how to do it:

1. We will go to the Cloud SQL Instances page in Google Cloud Console with the following link: `https://console.cloud.google.com/sql/instances`.

2. If you have already created a 12-month free trial account with Google Cloud, you can type in your email and click on **Next**; otherwise, click on **Create account** to sign up for a new account:

Figure Appendix.1 – Google Cloud Sign in

3. The following step will be to fill in the password and press the **Next** button:

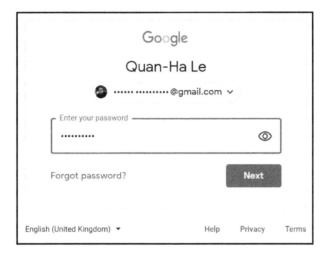

Figure Appendix.2 – Google Cloud Sign in (cont.)

4. In the following step, we will be in Google Cloud Platform and as it is our first time there, we will create our instance by clicking the **CREATE INSTANCE** label. Then we will go to the next step:

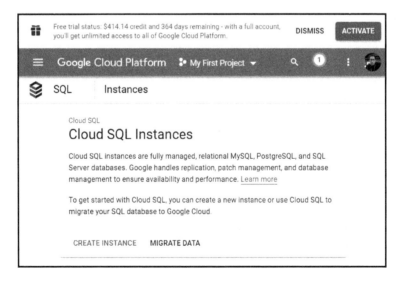

Figure Appendix.3 – Google Cloud SQL

5. The next step will be about creating our new project and we will choose the **NEW PROJECT** option in the modal window:

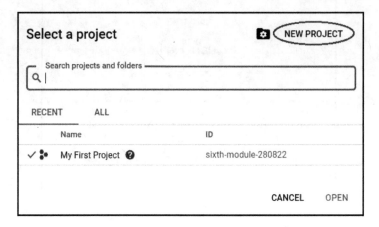

Figure Appendix.4 – Google Cloud Project

6. We will now give a name to our project, `Book Project`, and then press the **CREATE** button:

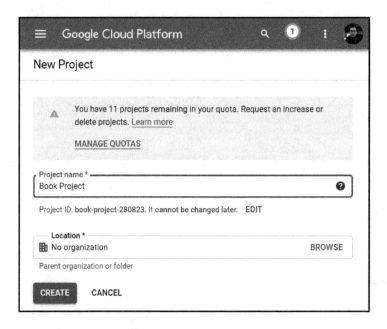

Figure Appendix.5 – Creating a new Google project

7. Google Cloud Platform will notify us to say our project has been created, as shown here:

Figure Appendix.6 – Project: Book Project is ready

8. Now that we have our project, it is time to create the first instance. We can do that using the **CREATE INSTANCE** button:

Figure Appendix.7 – Creating an instance within Book Project

9. The following step will be to select what kind of instance we want and we will choose PostgreSQL:

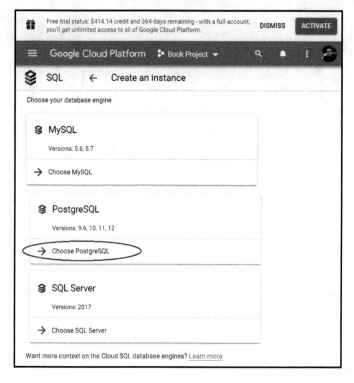

Figure Appendix.8 – PostgreSQL on Cloud SQL

10. After we have chosen PostgreSQL, Google Cloud Platform will process and create this new instance:

Figure Appendix.9 – PostgreSQL instance initialization

11. After the Compute Engine API has created our instance, in the following form, we will choose a couple of details:

- **Instance ID**: `atm-instance`.
- **Default user password**: `bookdemo`.

Use the default values for the other fields:

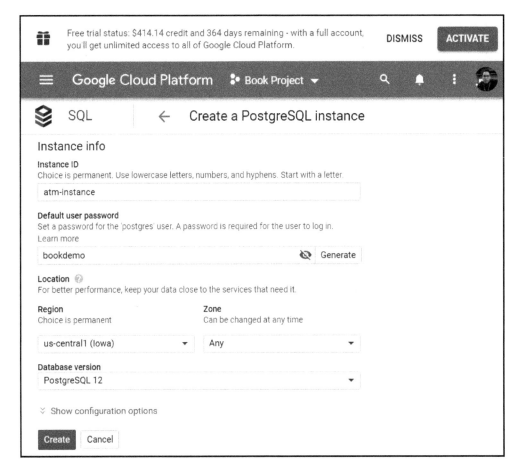

Figure Appendix.10 – Our PostgreSQL details on Google Cloud SQL

12. After that, we will press the **Create** button and the platform will notify us about instance creation:

Figure Appendix.11 – PostgreSQL instance progression

Instance creation could take a minute but the platform keeps us informed about it through an animation, as we can see in the following screenshot:

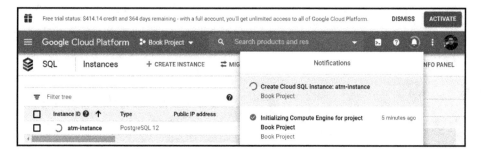

Figure Appendix.12 – PostgreSQL instance progression (cont.)

13. Meanwhile, if we return to the list of instances, we can see some important details that we will use in the future:

Figure Appendix.13 – PostgreSQL instance creation

We'll check out the following parameters here:

- **Instance connection name**: `book-project-280823:us-central1:atm-instance`
- **Public IP address**: `35.224.163.105`

14. When we will click on `atm-instance`, the platform will show another view where we will see options related to database instances:

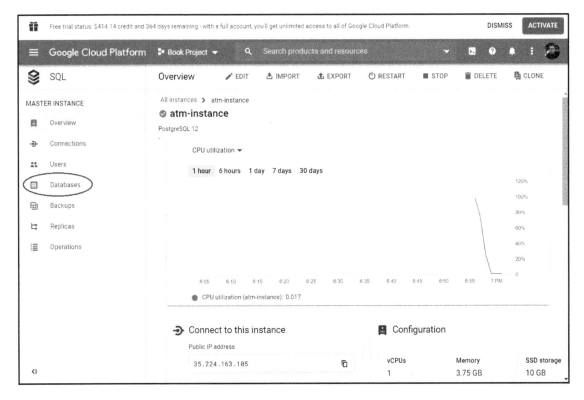

Figure Appendix.14 – PostgreSQL atm-instance dashboard

15. When we click on the **Databases** option, another view will be shown:

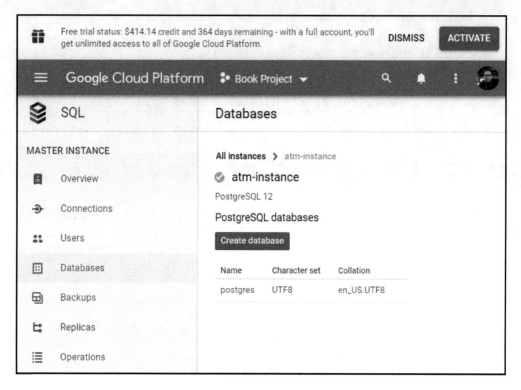

Figure Appendix.15 – Adding a database for the cloud instance

16. In this new view, we will click on the **Create database** button and it will require a name for it. We will choose `atm` as the name and we will click on **CREATE**:

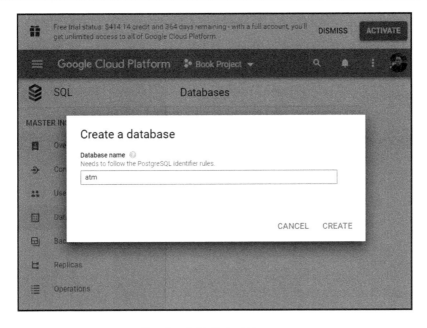

Figure Appendix.16 – Create a database screen

17. Afterward, the atm database will show up in the **Databases** view:

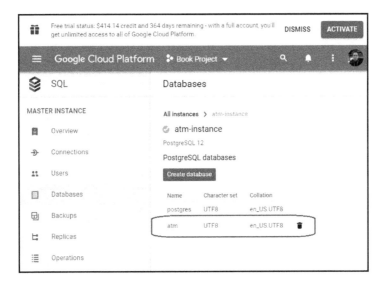

Figure Appendix.17 – The atm PostgreSQL database on Google Cloud SQL

18. Now it's time to configure **Connections**, and we will click on that label as follows:

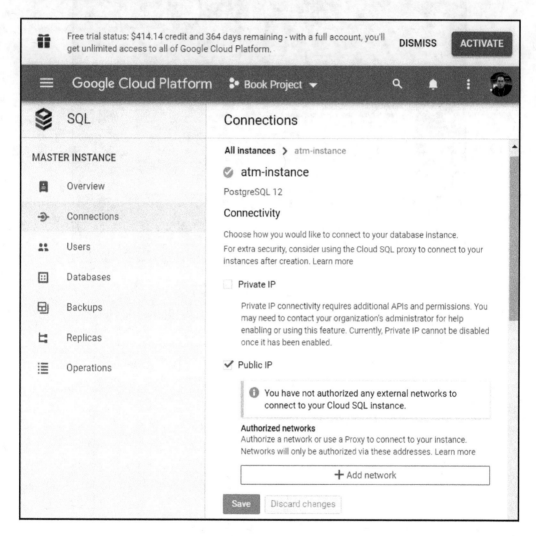

Figure Appendix.18 – Connection permission by Public IP

19. On the **Connections** view, we will click on the **Add network** button and we will add our current internet IP addresses:

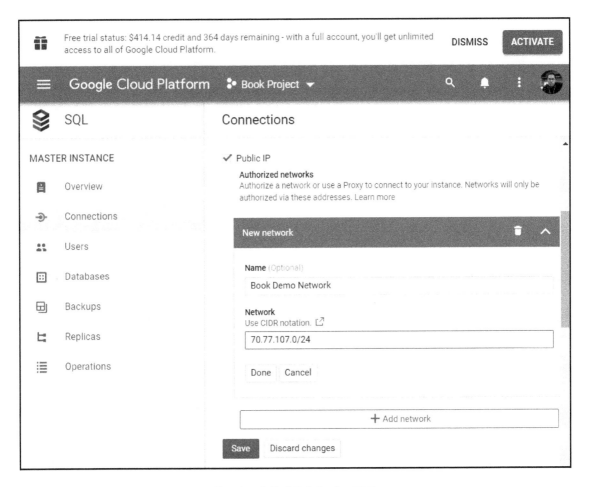

Figure Appendix.19 – Authorization of our network

We can use the `https://www.whatismyip.com/` service to find out our current IP. We will add that IP to the network.

20. Once the preceding configuration is done, we will see that our IP was allowed to access our database on Google Cloud Platform:

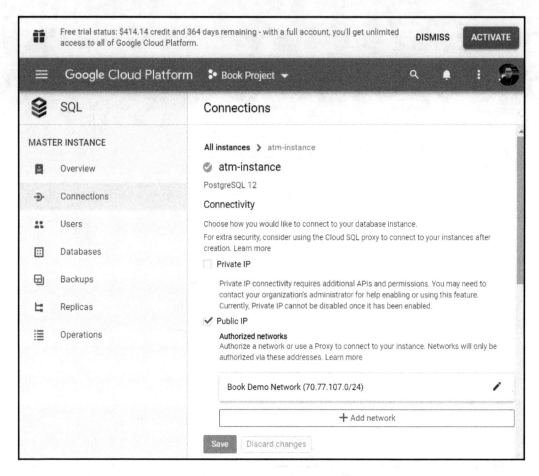

Figure Appendix.20 – Completion of our authorized networks

So far, we have guided you on how to follow the different steps to create a database in Google Cloud. In the next section, we will see how to connect to it through pgAdmin. Let's see how to do it.

pgAdmin connection for Google Cloud

A database in a cloud is nothing if we cannot access it, and that's why we will see how to connect to it through pgAdmin in the following steps:

1. On pgAdmin in our local environment, we will choose the **Create | Server...** option:

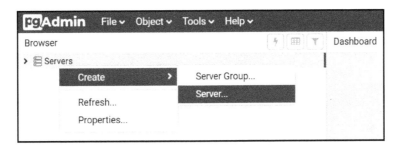

Figure Appendix.21 – pgAdmin for Google Cloud

2. After the previous step, the **Create-Server** view will be shown and there we will write a name to our connection on the **General** tab:

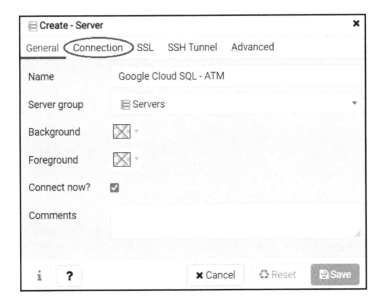

Figure Appendix.22 – Create-Server of Google Cloud

3. The following tab is **Connection** and there we will write the IP address that Google Cloud has provided in the previous section. After having entered the data as follows, we will save all this with the **Save** button:

- **Host name/address**: Google Cloud Instance public IP: 35.224.163.105
- **Maintenance database**: atm
- **Username**: postgres
- **Password**: bookdemo:

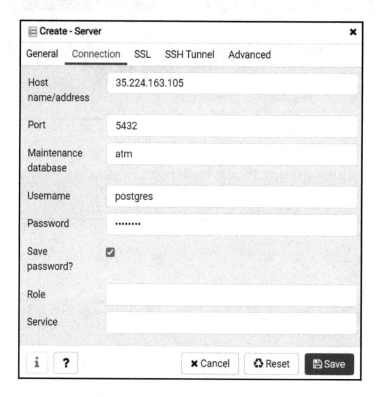

Figure Appendix.23 – Copying details of Google Cloud SQL

4. If everything went well, we should be able to expand **Google Cloud SQL - ATM**, and there we will be able to see that our **atm** database is ready to work with:

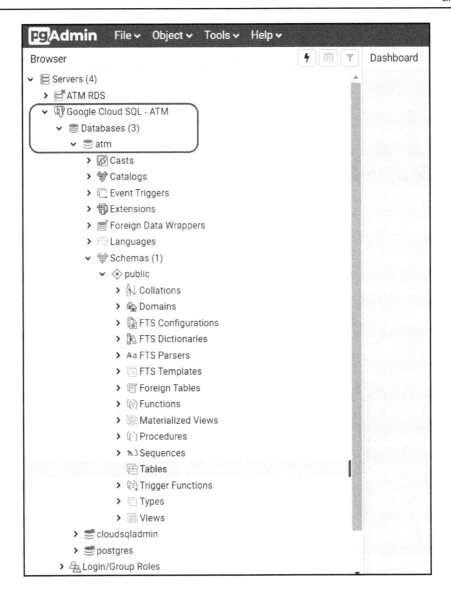

Figure Appendix.24 – Google Cloud SQL expansion

Now we can refer to the *Connecting to a PostgreSQL DB instance* section in Chapter 2, *Setting Up a PostgreSQL RDS for ATMs*, from *Step 6* to *Step 11* on pgAdmin, to create the ATM locations table.

In the following section, we will see another important cloud service, provided by Microsoft, called Azure. Let's see how to configure it.

Microsoft Azure for PostgreSQL

Azure is a set of cloud services from Microsoft. With Azure, it is possible to store information and create, manage, and deploy applications in the cloud. Among the different services present in Azure, we are going to work with databases and, more precisely, with PostgreSQL. Let's see how:

1. First of all, through the following link, we are going to register for an Azure account: `https://portal.azure.com/`.

2. We will enter our credentials to sign in to the portal:

Figure Appendix.25 – Azure portal Sign in

3. After the previous steps, we will see the default services dashboard view:

Figure Appendix.26 – Azure services dashboard

So far, we have seen how easy it is to connect with the services that Azure offers. In the next section, we will see how to create our PostgreSQL service in the cloud.

Creating an Azure database for the PostgreSQL server

The basis of any service in Azure is a resource. To create one, we are simply going to take the following steps:

1. We will create a resource by clicking the plus icon (+) and selecting **SQL databases**:

Figure Appendix.27 – Create a resource

2. Afterward, a list of databases will be shown and we will select **Azure Database for PostgreSQL**:

Figure Appendix.28 – Azure Database for PostgreSQL

3. Once the database of our preference has been selected, we will click on the **Create** button under the **Single server** option:

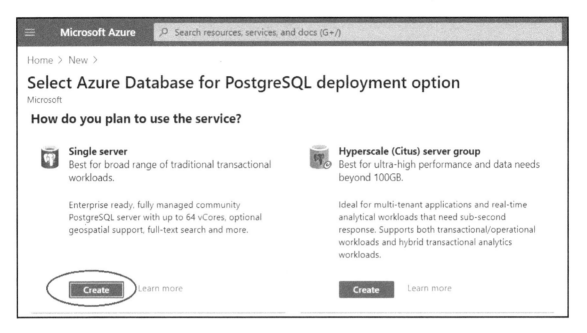

Figure Appendix.29 – Creating a single server

4. Now we will add a couple of parameter values to the deployment view:

- **Server name**: atm-server
- **Version**: 11
- **Admin username**: dba

- **Password:** Bookdemo20
- **Confirm password:** Bookdemo20:

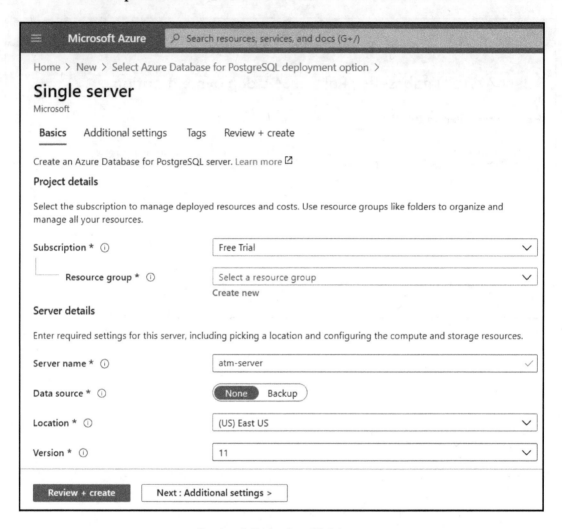

Figure Appendix.30 – Azure PostgreSQL deployment

5. We will click on the **Create new** link under **Resource group** and we will set the resource name as book-resources. After that, we will click on the **OK** button:

Figure Appendix.31 – Azure PostgreSQL deployment (cont.)

6. Now we will move to the **Review + create** tab and we will check the information:

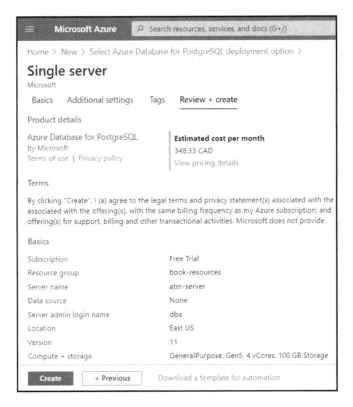

Figure Appendix.32 – Creating an Azure database for PostgreSQL

7. We will click on the **Create** button and it will launch the deployment process. It will take a few minutes:

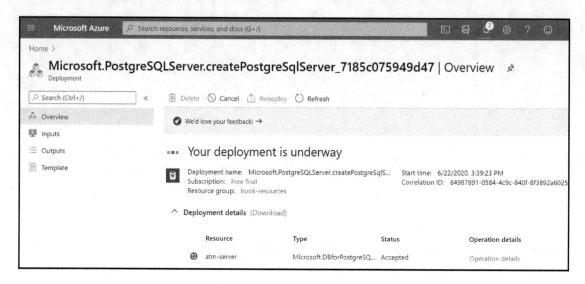

Figure Appendix.33 – Azure PostgreSQL deployment progression

8. Once the deployment is completed, we will click on **atm-server** located under **Deployment details**:

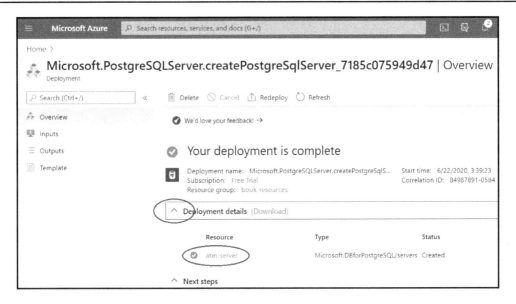

Figure Appendix.34 – Completion of Azure PostgreSQL deployment

9. Now we are in the **atm-server** view and we will click on **Connection security** in the **Settings** section on the left-hand side:

Figure Appendix.35 – Connection permission for Azure PostgreSQL

10. After the previous step, we will click on the **+ Add current client IP address** link and add any other IPs that we will use there. When we finish, we will click on **Save** in order for changes to be applied:

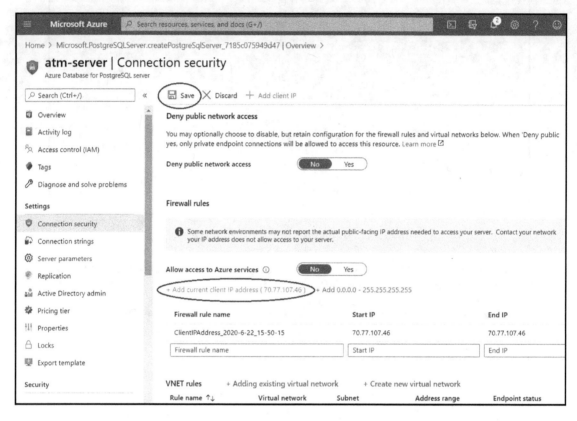

Figure Appendix.36 – Connection permission for Azure PostgreSQL (cont.)

So far, we have seen how to configure our resources in Azure. In this case, it is a PostgreSQL database but as we saw at the beginning of this section, there are many more, and with just a couple of clicks they will be available to us. Now we will see how to connect the database created previously to our local pgAdmin.

Getting an Azure resource connection for pgAdmin

Creating a resource in the cloud is useless if we don't use it. Let's see how to use the one previously created through pgAdmin:

1. We will click on **Overview** at the top on the left-hand side in our Azure database for the PostgreSQL server:

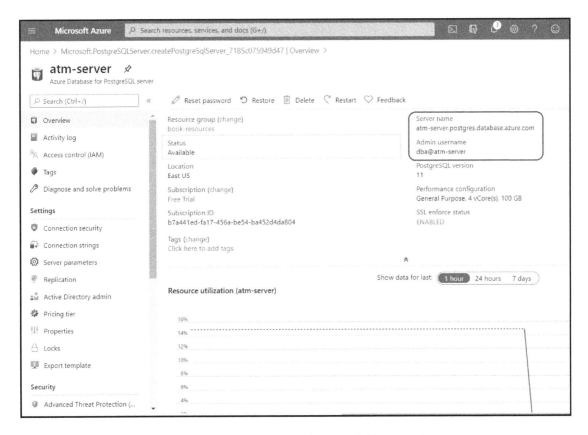

Figure Appendix.37 – PostgreSQL server credentials

From the preceding view, we will copy the following credentials:

- **Server name**: **atm-server.postgres.database.azure.com**
- **Admin username**: **dba@atm-server**

2. Now, in our local pgAdmin, we will click on **Create | Server...**:

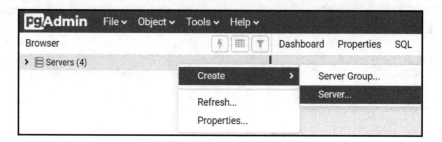

Figure Appendix.38 – pgAdmin for Azure

3. On the **Connection** tab, we will name our server `Azure PostgreSQL database`:

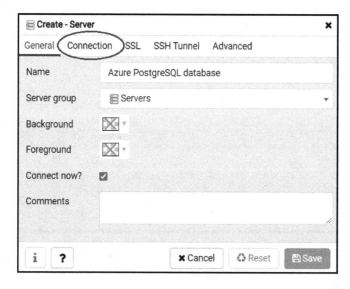

Figure Appendix.39 – Azure PostgreSQL server setup

4. The most important step is in the **Connection** tab, where we will add those parameters that we obtained in *Step 1* and then click on the **Save** button:

Figure Appendix.40 – Connection permission for Azure PostgreSQL (cont.)

5. Afterward, we will expand the Azure PostgreSQL database with pgAdmin and we will see our instance that was created in the previous section:

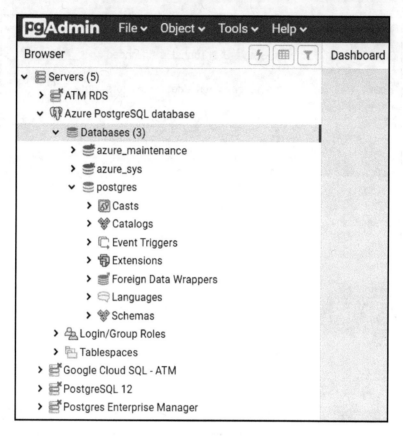

Figure Appendix.41 – Azure PostgreSQL expansion

6. Then, on the Azure PostgreSQL instance, we will right-click and create our `atm` database, as follows:

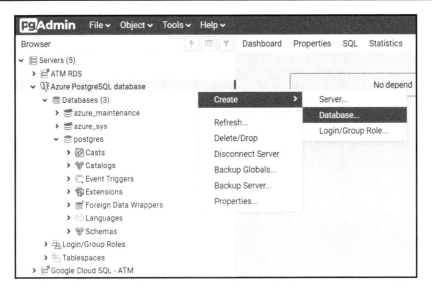

Figure Appendix.42 – Creation of the atm database

7. On the **Create-Database** window, and over on the **General** tab, we will add the following parameters:

- **Database**: atm
- **Owner**: dba

  After that, we will click on the **Save** button in order to finish our database creation:

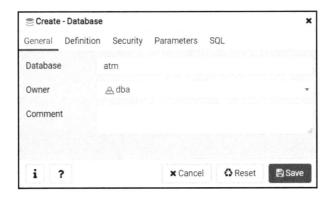

Figure Appendix.43 – Creation of the atm database (cont.)

8. Now it's time to expand the new `atm` database as follows. It is worth remembering that this is being recorded in our Azure cloud resource:

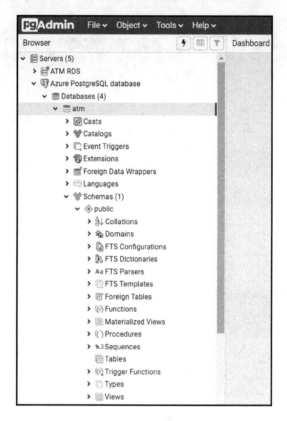

Figure Appendix.44 – Completion of the Azure atm database

Now we can refer to the *Connecting to a PostgreSQL DB instance* section, from *Step 6* to *Step 11* on pgAdmin, to create the ATM locations table for Azure.

So far, we have had the chance to see how easy it is to configure an Azure service in the cloud. However, it is important to mention that the up and down speeds provided by cloud services will always be lower than those provided in an on-premises environment, but perhaps customers will be willing to give up a bit of speed in exchange for the security of their data and other benefits as well.

We'll now check out another cloud service called Heroku.

Heroku Postgres

Heroku Postgres is a Heroku cloud database service based on PostgreSQL. Heroku Postgres provides features such as continuous protection, rollback, and high availability. It also allows for forking the database and creating followers and data clips.

Creating a Heroku app

Unlike the services we have seen previously, Heroku is a type of cloud service known as **Platform as a Service** (**PaaS**). In other words, this type of service is designed for the development of applications for companies of all sizes and helps to execute, implement, manage, and scale applications using various available computer programming languages. So, to create our Heroku Postgres, first we are going to create our application. Let's see how to do it:

1. As with all internet services, first, we will enter our Heroku credentials at the following link: `https://dashboard.heroku.com/apps`:

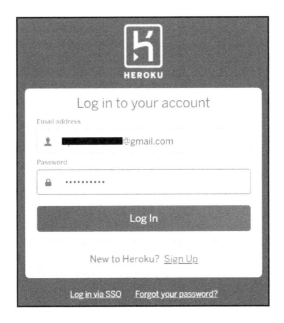

Figure Appendix.45 – Heroku dashboard login

2. After signing in, the Heroku dashboard is shown as follows and there we will click on **Create new app**:

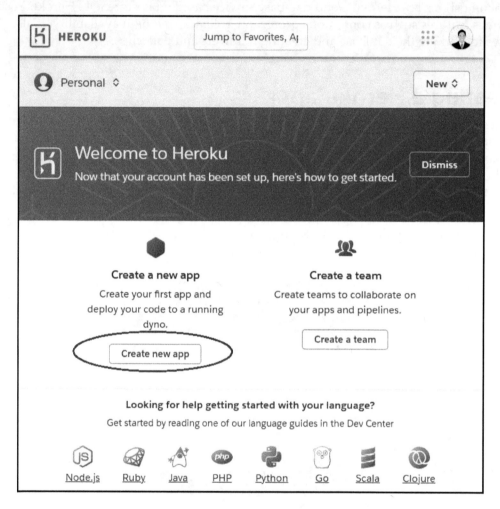

Figure Appendix.46 – Heroku dashboard

3. In the **Create New App** view, we will add `atm-heroku-app` as the app name and we will click on the **Create app** button:

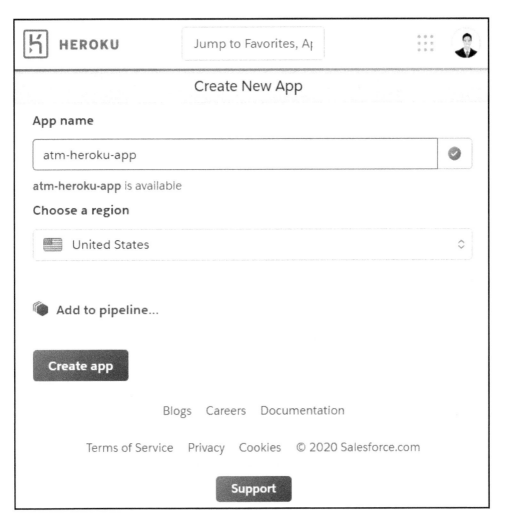

Figure Appendix.47 – Creating a new Heroku app

After we create the app, this is what we'll see:

Figure Appendix.48 – Creating a new Heroku app (cont.)

So far, we have taken the first steps to creating our database. In the next section, we will develop the **Resources** tab for this purpose. Let's see how to do it.

Creating a Heroku PostgreSQL database

Now, we already have our Heroku app and from here we will create our Heroku PostgreSQL. Let's see how to do it:

1. On the **Resources** tab, under **Add-ons**, we will search for Heroku Postgres and then we will select it from the suggested list:

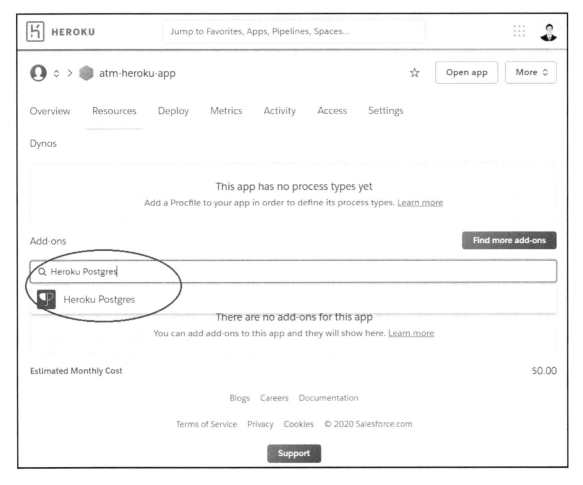

Figure Appendix.49 – Heroku Postgres add-on

2. Automatically, a pop-up window will show and there we will select **Hobby Dev - Free** for **Plan name**. Following this, we will click on the **Provision** button:

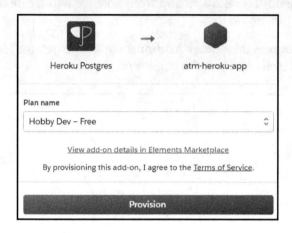

Figure Appendix.50 – Creation of Heroku Postgres

3. After we click the **Provision** button, we will see our Heroku Postgres plan and we will click on it:

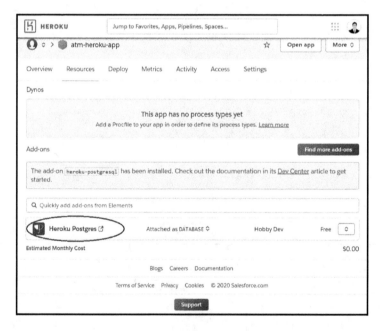

Figure Appendix.51 – Completion of Heroku Postgres add-on installation

4. Now, in the **Datastores | Heroku PostgreSQL** view, we will click on the **Settings** tab:

Figure Appendix.52 – Heroku Postgres settings

5. On the **Settings** tab, we will click on **View Credentials...**:

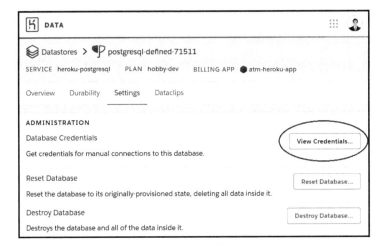

Figure Appendix.53 – Heroku Postgres settings (cont.)

6. On **View Credentials**, we will obtain all the data that we will need to connect from a client:

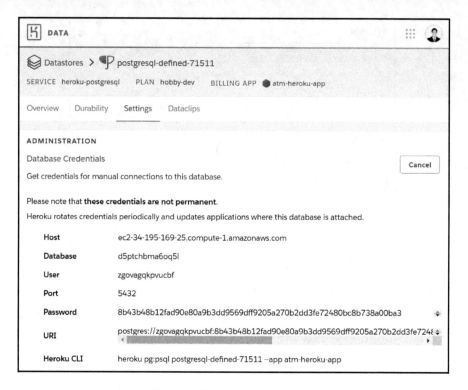

Figure Appendix.54 – Heroku Postgres credentials

The credential details here are as follows:

- **Host**: ec2-34-195-169-25.compute-1.amazonaws.com
- **Database**: d5ptchbma6oq5l
- **User**: zgovagqkpvucbf
- **Port**: 5432
- **Password**: 8b43b48b12fad90e80a9b3dd9569dff9205a270b2dd3fe72480bc8b738a00ba3

At this point, we already have our PostgreSQL database created in Heroku and the necessary data to connect from a client. As has been the case so far, our client for this purpose is pgAdmin, and in the next section, we will see how to connect from it.

Connecting Heroku PostgreSQL using pgAdmin

As we previously mentioned, now is the time to connect from our local location to our Heroku service through pgAdmin. Let's see how to do it:

1. On our local pgAdmin in the browser, we will right-click on **Servers** and then click on **Create | Server...**:

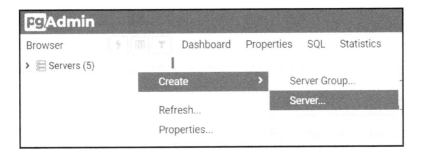

Figure Appendix.55 – pgAdmin connection for Heroku

2. On the **Connection** tab, we will name it Heroku Postgres:

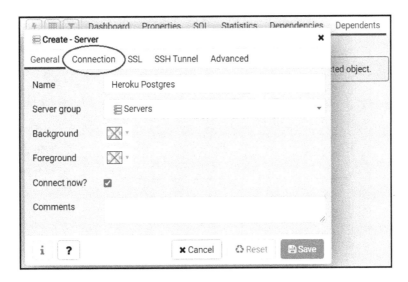

Figure Appendix.56 – Heroku Postgres server setup on pgAdmin

3. Following that, we will add our credentials previously obtained on Heroku on the **Connection** tab:

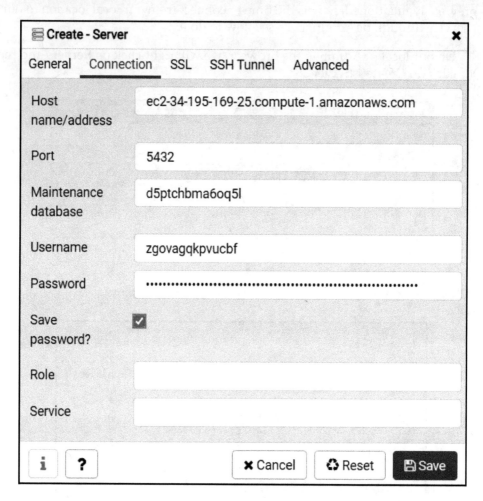

Figure Appendix.57 – Heroku Postgres server setup on pgAdmin (cont.)

4. Once the credentials have been validated, we will expand the Heroku Postgres instance and over a public schema, we will right-click and select **Query Tool** from the contextual menu:

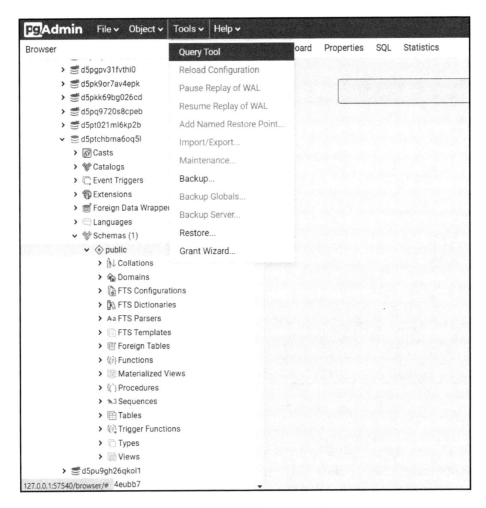

Figure Appendix.58 – Heroku Postgres server expansion

5. On the new **Query Editor** tab, we can refer to Chapter 2, *Setting Up a PostgreSQL RDS for ATMs*, in the *Connecting to a PostgreSQL DB instance* section, from *Step 6* to *Step 11* on pgAdmin, to create the ATM locations table for Heroku Postgres:

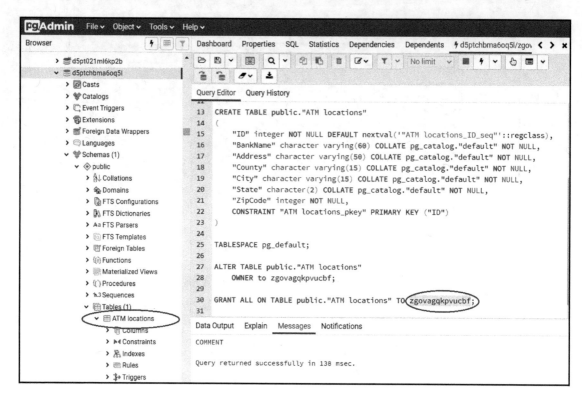

Figure Appendix.59 – Heroku Postgres ATM locations table

However, as Heroku Postgres does not support it, we cannot rename the database name `atm` nor can we create it as or rename the username `dba`; therefore, when we re-execute the steps from Chapter 2, *Setting Up a PostgreSQL RDS for ATMs* on pgAdmin, we will have to fix the username of its GitHub script from `dba` to `zgovagqkpvucbf` (Heroku user) as follows:

```
CREATE SEQUENCE public."ATM locations_ID_seq"
    INCREMENT 1
    START 658
    MINVALUE 1
    MAXVALUE 2147483647
    CACHE 1;

ALTER SEQUENCE public."ATM locations_ID_seq"
```

```
    OWNER TO zgovagqkpvucbf;

GRANT ALL ON SEQUENCE public."ATM locations_ID_seq" TO zgovagqkpvucbf;

CREATE TABLE public."ATM locations"
(
    "ID" integer NOT NULL DEFAULT nextval('"ATM
locations_ID_seq"'::regclass),
    "BankName" character varying(60) COLLATE pg_catalog."default" NOT NULL,
    "Address" character varying(50) COLLATE pg_catalog."default" NOT NULL,
    "County" character varying(15) COLLATE pg_catalog."default" NOT NULL,
    "City" character varying(15) COLLATE pg_catalog."default" NOT NULL,
    "State" character(2) COLLATE pg_catalog."default" NOT NULL,
    "ZipCode" integer NOT NULL,
    CONSTRAINT "ATM locations_pkey" PRIMARY KEY ("ID")
)

TABLESPACE pg_default;

ALTER TABLE public."ATM locations"
    OWNER to zgovagqkpvucbf;

GRANT ALL ON TABLE public."ATM locations" TO zgovagqkpvucbf;

COMMENT ON TABLE public."ATM locations"
    IS 'ATM locations of New York city';
```

As in the preceding screenshot, you can see that we have successfully created the ATM locations table with pgAdmin for Heroku Postgres.

As we said at the beginning, Heroku is a cloud service designed mainly to implement applications in the cloud, and perhaps due to this there are some limitations that we found with PostgreSQL. However, it is worth taking it into account as a service since many have been doing so and giving good reviews for it.

We will now check out a natively Postgres database service called EnterpriseDB.

EnterpriseDB cloud database

In order to finalize our study on the different offers that exist in the DBaaS market, now it is time to show one of the pioneers of which PostgreSQL is concerned. Maybe if you are looking for something natively PostgreSQL, EDB would be the answer. In the following sections, we will address the creation of a PostgreSQL database and its access from a client as we did with the previous vendors.

Creating a PostgreSQL cluster

One of the many services that EDB offers in the cloud is the creation of a PostgreSQL instance. Let's see how to do this in the following steps:

1. We will go to the following link: `https://www.enterprisedb.com/edb-postgres-cloud-database-service`.

2. We will add our credentials to sign in to the EDB cloud database; otherwise, you will need to sign up for a new account there through the next link:

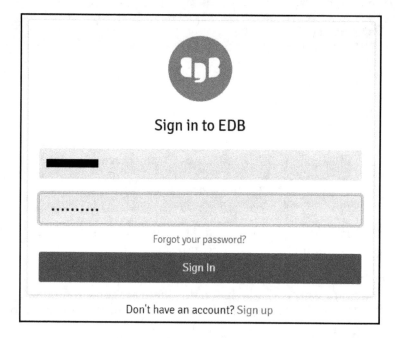

Figure Appendix.60 – EDB cloud Sign in

3. If this is the first time we are signing in, we will need to select the free trial by clicking the **Start free** button:

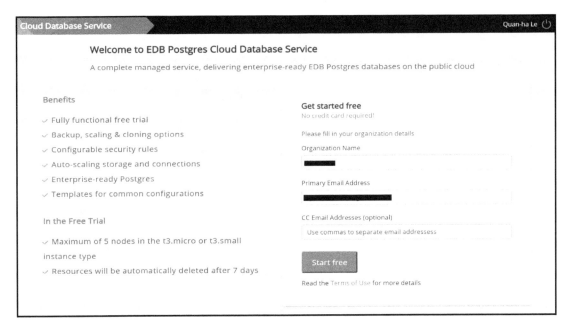

Figure Appendix.61 – EDB cloud free trial

4. Once we have logged in to the cloud database service, we will select the **US East** zone and it will pop up a new cluster tab:

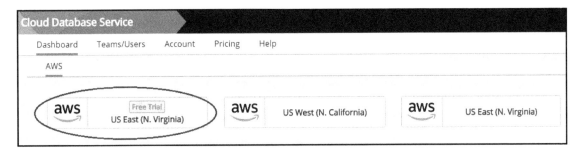

Figure Appendix.62 – Cloud Database Service

5. In this new view, we will click on the **New Cluster** icon on the left-hand side:

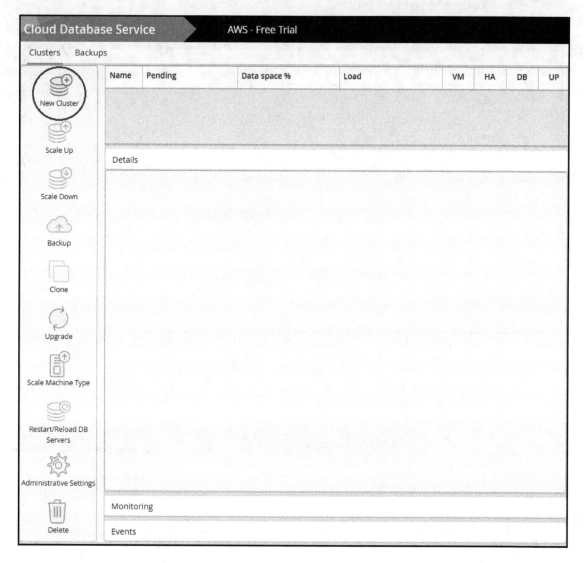

Figure Appendix.63 – Cloud Database Service (cont.)

6. The previous action will open a new window popup where we will select **Launch from Template**:

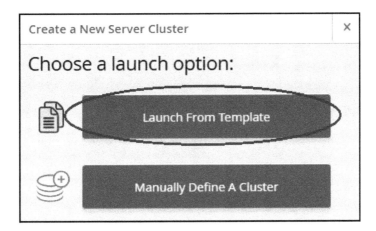

Figure Appendix.64 – Launching a new PostgreSQL cluster

7. A new view will open and there we will add the following parameters:

- **PostgreSQL**: Yes (checked)
- **Select Version(s)**: **12**
- **Cluster Name**: atm-cluster
- **DB Master User**: clouddba (because they require 4-20 characters)

- **DB Master Password**: bookdemo:

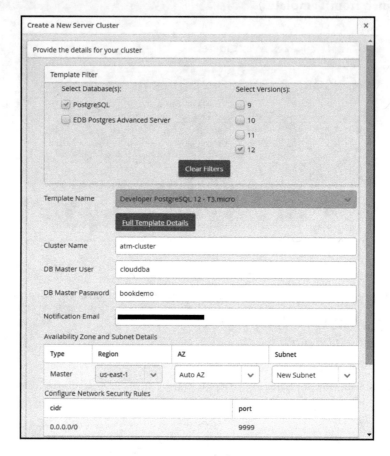

Figure Appendix.65 – PostgreSQL cluster setup

In the same view, we will delete the existing rule:

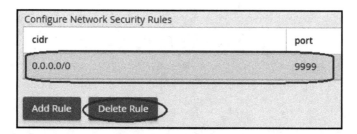

Figure Appendix.66 – PostgreSQL cluster network security

We will need to confirm the **Delete Rule** dialog by clicking on **Delete**:

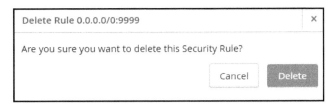

Figure Appendix.67 – PostgreSQL cluster network security (cont.)

8. Now we will add a new rule and we will click on the **Add Rule** button:

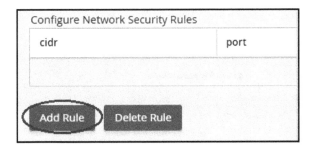

Figure Appendix.68 – PostgreSQL cluster network security (cont.)

9. On the popping up of the **Add Rule** dialog, we will add our CIDR. As we did before, we can use `https://www.whatismyip.com` to obtain our IP address and we will click on the **Apply** button:

Figure Appendix.69 – PostgreSQL cluster network security (cont.)

The new rule will be shown as follows and we will click on the **Launch** button:

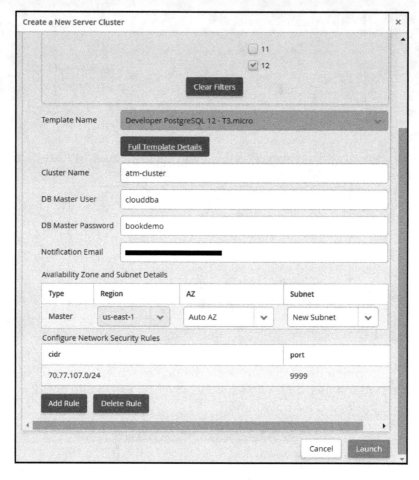

Figure Appendix.70 – Completion of PostgreSQL cluster network security

12. A pop-up window will show us that the creation process has begun as we will see in the following screenshot:

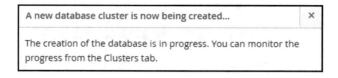

Figure Appendix.71 – Database creation progression

13. After a few minutes, when the cluster has been created, we will click on **atm-cluster** and we will copy the details from there:

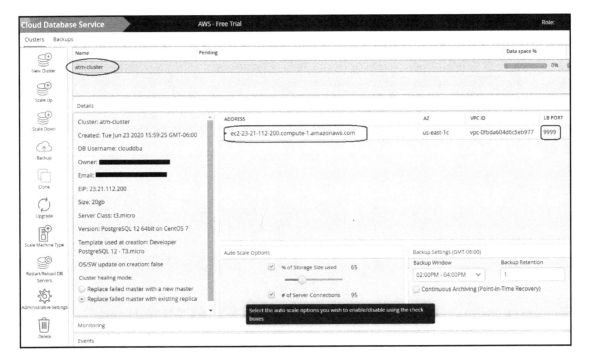

Figure Appendix.72 – Completion of atm-cluster creation

We'll check out the following parameters here:

- **ADDRESS**: `ec2-23-21-112-200.compute-1.amazonaws.com`
- **LB PORT**: `9999`

14. At the same time, we will receive a notification email as follows where all the credentials are provided:

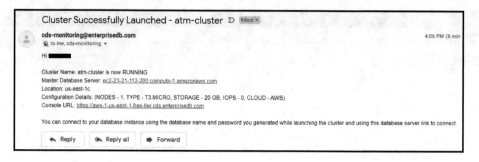

Figure Appendix.73 – Notification of cluster creation

So far, we have seen that the steps were similar to those we have done with other vendors, except that here we are only talking about PostgreSQL; we do not have other database services. In the next section, we will see how to connect to this service created through pgAdmin.

Using pgAdmin to connect to EDB PostgreSQL

The following steps are already known, so we will go directly on to adding the necessary details to connect:

1. We assume that previously the new server has already been created on pgAdmin, so now we will go to the **General** tab and we will set the name to EDB PostgreSQL - ATM:

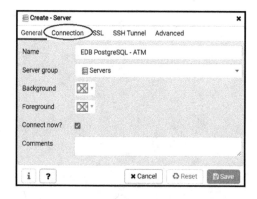

Figure Appendix.74 – EDB PostgreSQL server setup on pgAdmin

2. In the **Connection** tab, we will add the credentials that we received in the preceding email from EDB:

- **Host name/address**: `ec2-23-21-112-200.compute-1.amazonaws.com`
- **Port**: `9999`
- **Username**: `clouddba`
- **Password**: `bookdemo`
- **Save password?**: Yes (checked):

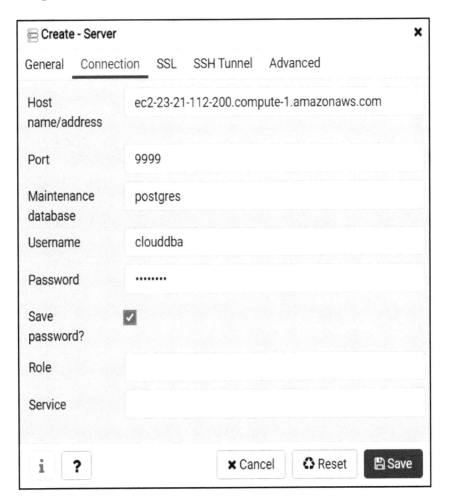

Figure Appendix.75 – EDB PostgreSQL server setup on pgAdmin (cont.)

3. Then we will be able to expand the **EDB PostgreSQL - ATM** server with pgAdmin:

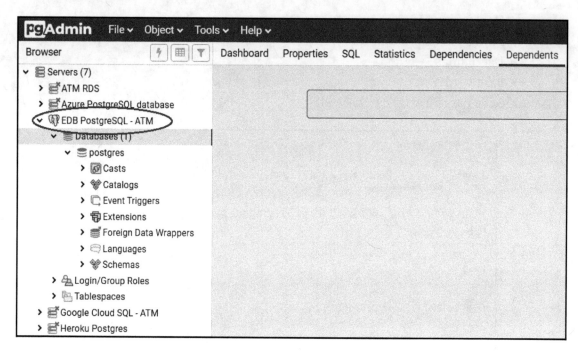

Figure Appendix.76 – EDB PostgreSQL - ATM expansion

4. Now, on **EDB PostgreSQL - ATM**, we will create a new `atm` database:

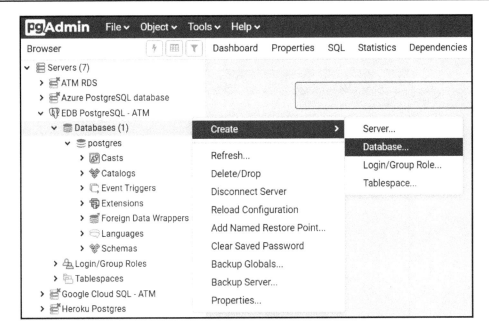

Figure Appendix.77 – Creating the EDB PostgreSQL atm database

5. In the **Create | Database...** window, we will add the following parameters and save them:

- **Database**: atm
- **Owner**: clouddba:

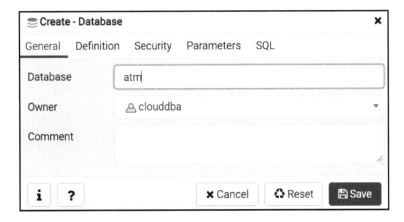

Figure Appendix.78 – Creating the EDB PostgreSQL atm database (cont.)

6. Now we will be able to expand the newly created `atm` database, as shown in the following screenshot:

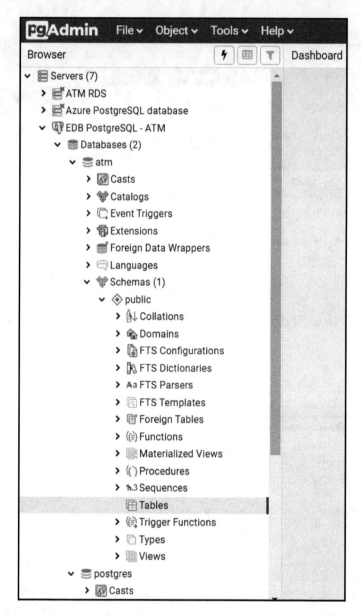

Figure Appendix.79 – Completion of the EDB PostgreSQL atm database

8. Now we can refer to Chapter 2, *Setting Up a PostgreSQL RDS for ATMs*, in the *Connecting to a PostgreSQL DB instance* section, from *Step 6* to *Step 11* on pgAdmin, to create the ATM locations table for EDB PostgreSQL; however, our GitHub is always using `username = dba`, we will use a new `username = dba` for PGAdmin so that all of our GitHub scripts can easily be executed.

For that, we will select the **Tools** menu | **Query Tool** and we will execute the following statement:

```
CREATE USER dba WITH ENCRYPTED PASSWORD 'bookdemo';
GRANT ALL PRIVILEGES ON DATABASE atm TO dba;
```

This will result in the following output:

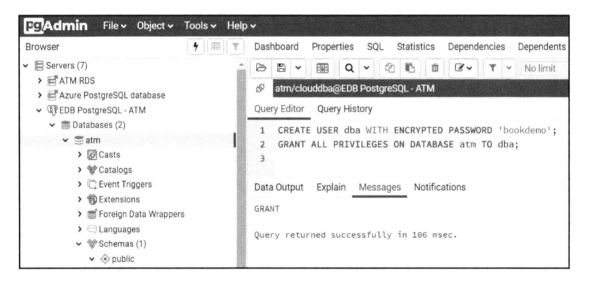

Figure Appendix.80 – Creating the dba user for the EDB PostgreSQL atm database

Hence, now, we can re-apply the ATM locations table through EDB PostgreSQL in the cloud.

So far, we have seen that connecting to the EDB cloud is similar to other, previous services. The important thing is to always have the credentials at hand, and that's it.

Summary

In this *Appendix*, we have learned DBaaS skills such as how to set up and manage PostgreSQL step by step with Google Cloud SQL, Azure, Heroku, and the EDB cloud. Keep in mind that the cloud providers seen are the most used and best-ranked for the DBaaS application.

For your future work after you complete this book, you can develop your own research and technology deployment from your PostgreSQL projects. Whenever you discover something new, you can contribute your invention to the PostgreSQL association with our network of PostgreSQL conferences all over Canada and the United States by contacting the Diversity Committee of the United States PostgreSQL Association here: `https://postgresql.us/diversity/`.

Thank you so much for your attention throughout the book!

Other Books You May Enjoy

If you enjoyed this book, you may be interested in these other books by Packt:

Learn PostgreSQL
Luca Ferrari, Enrico Pirozzi

ISBN: 978-1-83898-528-8

Learn how to clean your data and ready it for analysis

- Understand how users and connections are managed by running a PostgreSQL instance
- Interact with transaction boundaries using server-side programming
- Identify bottlenecks to maintain your database efficiently
- Create and manage extensions to add new functionalities to your cluster
- Choose the best index type for each situation
- Use online tools to set up a memory configuration that will suit most databases
- Explore how Postgres can be used in multi-instance environments to provide high-availability, redundancy, and scalability

PostgreSQL 13 Cookbook
Vallarapu Naga Avinash Kumar

ISBN: 978-1-83864-813-8

- Understand logical and physical backups in Postgres
- Demonstrate the different types of replication methods possible with PostgreSQL today
- Set up a high availability cluster that provides seamless automatic failover for applications
- Secure a PostgreSQL encryption through authentication, authorization, and auditing
- Analyze the live and historic activity of a PostgreSQL server
- Understand how to monitor critical services in Postgres 13
- Manage maintenance activities and performance tuning of a PostgreSQL cluster

Leave a review - let other readers know what you think

Please share your thoughts on this book with others by leaving a review on the site that you bought it from. If you purchased the book from Amazon, please leave us an honest review on this book's Amazon page. This is vital so that other potential readers can see and use your unbiased opinion to make purchasing decisions, we can understand what our customers think about our products, and our authors can see your feedback on the title that they have worked with Packt to create. It will only take a few minutes of your time, but is valuable to other potential customers, our authors, and Packt. Thank you!

Index

www.ingramcontent.com/pod-product-compliance
Lightning Source LLC
LaVergne TN
LVHW081328050326
832903LV00024B/1071